Authors of the Storm

Authors of the Storm

Meteorologists and the Culture of Prediction

Gary Alan Fine

The University of Chicago Press :: Chicago and London

The University of Chicago Press, Chicago 60637
The University of Chicago Press, Ltd., London

Paperback edition 2010
Printed in the United States of America

19 18 17 16 15 14 13 12 11 10   2 3 4 5 6

ISBN-13: 978-0-226-24952-0 (cloth)
ISBN-13: 978-0-226-24953-7 (paper)
ISBN-10: 0-226-24952-2 (cloth)
ISBN-10: 0-226-24953-0 (paper)

Library of Congress Cataloging-in-Publication Data

Fine, Gary Alan.
Authors of the storm : meteorologists and the culture of prediction / Gary Alan Fine.
p. cm.
Includes bibliographical references and index.
ISBN-13: 978-0-226-24952-0 (cloth : alk. paper)
ISBN-10: 0-226-24952-2 (cloth : alk. paper)
1. Meteorologists—Psychology. 2. Weather forecasting—Methodology. 3. Meteorology—Social aspects. I. Title.
QC869.5.F56 2007
551.63—dc22

2006025445

∞ The paper used in this publication meets the minimum requirements of the American National Standard for Information Sciences—Permanence of Paper for Printed Library Materials, ANSI Z39.48-1992.

To Peter Gregory Fine
whose dynamics are always fluid

Contents

Preface

Everybody talks about the weather. It is one of the shared interests that bind us together, building our social capital and contributing to the integration of civil society. Talk of weather is not limited by age, race, gender, class—the core demographic variables that divide us. Unlike politics or religion—two difficult topics for mere acquaintances—we assume that everyone is competent to talk about the weather and that we share attitudes and emotions toward meteorological events. The skies provide common ground. Even when this assumption doesn't hold, disagreements are not taken as grave affronts. They simply further the conversation. Public fascination with weather is equally evident in media, not only the startlingly successful Weather Channel, but popular films such as *Twister* and *The Day after Tomorrow*.[1]

Yet, weather as a human concern goes beyond mere talk; it affects how we feel and how we live. Weather shapes our behaviors, determines the material conditions of our lives (our clothing, housing, and health), and channels our emotional well-being. As weather becomes climate—long-term atmospheric patterns—our sense of self, occupational order, and even our social structure are molded. National or regional character is sometimes attributed to the climatic conditions of society, as in alleged differences in character between Nordic and Mediterranean peoples, ice people and sun people.

: : :

I have long been fascinated by weather. Like many meteorologists, my interest was forged in childhood, as I watched temperatures, precipitation, and storm systems march uncertainly across television weather maps. As forecasters speak of their early interest, I was a toddler meteorologist. Earth Science—what a charming label for a science!—was my favorite science class. This research project permitted my return to a youthful fascination. I now understand it not only as natural science, but as social science.

In addition to satisfying my childhood preoccupation, this volume contributes to several domains of scholarship including the ethnography of laboratory life,[2] small group dynamics, the sociology of work, and the sociology of culture. Weather offices, of course, are not precisely laboratories, and meteorology is not a bench science, although what and where laboratories are has become a topic of debate.[3] These offices are closely linked to what Bruno Latour has described as "centers of calculation."[4] The widely quoted saying of Mark Twain with which I began this preface ends, "but nobody does anything about it." Operational meteorology (specifically weather forecasting) is a predictive arena, but it is not, yet, an applied science or an experimental science. It is, nonetheless, what I label a "public science," a discipline whose conclusions are presented in plain public view, with citizens as the primary consumers of their claims, along with such specialized occupational workers as pilots, contractors, and gardeners. Operational meteorology, like horticulture, ecology, and public health, bridges science and public service.

The workers with whom I spent time are meteorologists employed by the National Weather Service. Their primary task is not to change the weather[5] nor to describe it, but to forecast its course. Their responsibility is to share the future with their fellow citizens[6] and with influential organizations and corporations. By colonizing the future,[7] they shape our approaches to risk management as well as our routines of life. While my ethnography contributes to understanding public science, it also depicts a work world that, because of its focus on forecasting the future, may have more in common with stock analysis or medical practice than with industrial chemistry or high-energy physics.

Meteorologists' predictions are produced in offices filled with equipment and within a national and global technological system by which information is gathered. These workers have the challenge of maintaining their autonomy, professionalism, and authority while relying on technological measurements of weather in the context of organizational demands to produce timely forecasts. Put another way, they must convince their public that they are more than machine-readers, that they have authentic experience that generates trust. Forecasters examine meteorological signs in the intimate cocoon of the present, predicting what will happen in the next few minutes and the next ten days.

Some public talk about the weather gleefully refers to the challenges of getting forecasts right. That forecasts are often incorrect in large or small ways gives us license to complain that weather forecasters always "get it wrong," a source of endless jokes. One should only wash one's car when the forecast is for rain; on that day dry conditions will surely prevail. Such a perspective leads some meteorologists to avoid publicly announcing their occupation. We—and they—recall the mistakes, while the many correct forecasts are taken for granted and forgotten.

I deeply appreciate the hospitality of the talented staff of the National Weather Service. The courtesy that I have been shown at every level of the organization has been remarkable, particularly given that I was never pressured to reach any conclusions. Of those who need not be anonymous, I thank General Jack Kelly, the director of the National Weather Service at the time of this research, Louis Uccellini, the head of the National Centers for Environmental Prediction in Camp Springs, Maryland, and Joseph Schaefer, the head of the Storm Prediction Center in Norman, Oklahoma.

I have deep fondness and admiration for Paul Dailey, now retired, who was the meteorologist-in-charge of the Chicago office at the time this research was conducted. I recall the first day that I was scheduled to meet with Mr. Dailey. I was nervous; he had the fate of my research in his hands, and on the drive to the office I rehearsed explanations that I might present to persuade him to permit me to observe his office. When I arrived, he asked, "When do you want to start?" All that anxiety for naught. Thank you, Paul, for that and for your decency, a trait which all of your staff treasured. My gratitude extends to the staff in the three local offices I studied, to the staff at the Storm Prediction Center, the National Severe Storms Laboratory, and the National Center for Environmental Prediction, particularly the Environmental Modeling Center. Although some of those being described may feel that particular details are embarrassing, hurtful, or incomplete, I never intended to write an expose, nor did I. These are admirable people, and I left with increased confidence in the judgments of workers in this government agency. My descriptions of the "underside of work" at the National Weather Service are comparable to accounts of many other occupations. Depictions of research universities reveal deviations from professional norms that are as dramatic.

Among my colleagues I particularly value the following for their support and advice: Ken Alder, Joseph Banas, Adele Clarke, Paul Dailey, Shripad (Jayant) Deo, Thomas Gieryn, Timothy Hallett, Pauline Kusiak, Ken Labas, Michael Lynch, Christena Nippert-Eng, Michael Sauter, Lucy Suchman, Casey Sullivan, and Diane Vaughan. Many of them provided detailed comments on this manuscript or on one of the articles that flowed from it. My writing began while I was a Fellow at the Swedish Collegium for Advanced Study in the Social Sciences in Uppsala, Sweden, during the winter and spring of 2003. If the weather was not always warm and sunny, the fellowship was. Final editorial changes were made while I was a fellow at the Russell Sage Foundation in New York City during the academic year 2005/6. For their help in transcribing the interviews from this research, I thank Deborah Nelson, Emily Long, and Jane Friedman. The index was prepared by Todd Fine.

I am grateful for permission to use previously published material in two chapters of this book. Part of chapter 2 first appeared as "Shopfloor Cultures: The Idioculture of Production in Operational Meteorology," *Sociological Quarterly* 47 (2006): 1–19, published by Blackwell Publishing. Part of chapter 5 first appeared as "Ground Truth: Verification Games in Operational Meteorology," *Journal of Contemporary Ethnography* 35 (2005): 1–20; copyright 2005, reprinted by permission of Sage Publications.

And, of course, my thanks to my family—my wife, Susan, and my (now adult) sons, Todd and Peter—for being supportive of a spouse and parent who was more eager to talk about the weather than was either common or decent.

Introduction

Remember, when the weather goes bad, we're in sales, not production.

National Weather Service employee

: : :

The question is not weather but when. Humans are exquisitely sensitive to their ambient environment. We are *Homo meteorologicus*. As Verlyn Klinkenborg puts it, weather is "one of the moral dimensions in which we live."[1] Weather shapes our choices of action. Being planners, we require guides for approaching weather systems. Whenever we desire technical control, we anoint ritual knowledge specialists. Sometimes we demand that the atmosphere be altered. This is the task of rainmakers. For modern societies, prediction is the aim. (Recent attempts to alter the atmosphere have not met with great success or public approval.)[2]

At the heart of weather as an occupational world are those specialists that the public and the state call upon to reveal the future. We ask these workers what to expect in temperature, precipitation, wind, and humidity. We label these workers weather forecasters, or, more formally, operational meteorologists. Most people have their most salient meteorological connection with media figures, charismatic individuals who may or may not have expertise in meteorology. Too often we ignore those individuals who stand behind these attractive, recognizable, friendly faces.

Prognoses of the future are created and shared, shaped by the contours of group life. Weather knowledge results from the practices of forecasters employed by the National Weather Service, the government agency whose responsibility and claimed expertise includes meteorological prediction.[3] In this, the group life and culture of meteorologists intersect with the organizational demands that are placed on them. I watched as they struggled to create authority in their work life, establishing a right to know. To see how weather forecasts are created and distributed is to understand some of the conditions of *public science,* the making of authoritative, systematic, and empirical claims of the natural environment to a popular audience.[4] In this focus on public science, meteorologists share similar concerns with pharmacists, public health officers, ecologists, agricultural agents, foresters, dieticians, horticulturalists, and even one's family physician.

To understand the doing of weather work, I draw on research from that branch of the social sciences known as science studies. An outsider, I incorporate concerns with worklife, group dynamics, and culture. The underlying claim of science studies is that science—and the knowledge claims that result—is socially organized, not merely a window into truth. That much is valid. Yet, despite our skepticism of confident certainty, scientific claims are based on empirical reality, even though that reality is understood through collective understandings. Some critics unfairly stereotype science studies as a brand of radical doubt and deconstructionism. But anyone who has had a medical procedure, has operated a computer, or has planned a picnic is grateful for the accomplishments of scientists, even when imperfect in their predictions and practices. The salient questions are how these workers come to make the claims they do, how they work to produce these claims, how they feel about those claims, and how they persuade audiences that their claims are plausible.

In this vein I describe how practitioners of one knowledge arena operate. Still, meteorology is not basic science, housed in a university where social control tends to be discreetly hidden. That these practitioners work for the United States government in small offices within a large bureaucracy is critical. In addition, meteorology has relatively low-status in the hierarchy of the sciences, an offspring of the marriage of fluid dynamics physics and earth sciences. None of America's elite Ivy League universities has a department that specializes in meteorology.[5] And the workers that I observed are not oriented to research. These are not professors or Ph.D.s, but, for the most part, men and women with Bachelor of Science degrees in meteorology, with some M.S. degree holders and a few who lack a college degree. Each local office (of which there are 122) produces two forecasts daily (around 3:30 p.m. and 4:00 a.m.) as well as updates (Nowcasts) for significant precipitation and warnings if severe weather threatens. Practices in local offices are shaped by rules promulgated by headquarters in Washington. I hesitate to term these men and women bureaucrats, given the stigma of the label, but they belong to a bureaucracy with strengths in rationalization and in uniformity of practice.[6]

Most weather forecasts are "good enough." Their accuracy is impressive for the following day's weather, and sufficient to rely upon for two or three days. (I will later have much to say about "accuracy"). These forecasters do more than throw darts at the wind.

Within the broader attempt to understand the conditions of work of these forecasters, I address a set of theoretical issues: (1) how weather forecasting gets done given the bureaucratic obstacles and temporal pressures under which meteorologists operate, (2) how science is defined by these workers and how they situate themselves in relationship to this honorific category, (3) how meteorologists and others depict future events and then justify that depiction, colonizing the future by building on a present and past that they have constructed by means of the machines that they employ, (4) how occupational autonomy maintains itself—in this case embedded in control of language and images—in the face of organizational and technological control, (5) how scientific, predictive accuracy is created through the organizational demands by which forecasts are verified, and (6) how the relationships between these workers and others who stand outside of the boundary of their workplace, including mass media, private companies, and government agents, shape meteorological practice. In each instance, scientific work and one's identity as a scientist are linked to authentic knowledge and legitimacy.

These are broad issues, with implications that extend to other spheres of social life. Meteorologists, despite their seemingly specialized and esoteric work, face many of the same challenges of autonomy, prediction, and evaluation as other workers.

A Student of the Air

Having claimed generalization, the craft of meteorology is also specialized and unique. This is an occupation that an outsider, a naive sociologist, with only a junior high school course in earth science and a high school course in physics could not fully master, no matter how many hours watching the Weather Channel and the antics of Al Roker and Willard Scott. At best, in the terms of Harry Collins and Robert Evans,[7] I acquired *interactive expertise,* able to converse about basic meteorological concerns and to translate them to others.

To conduct an ethnography of a technical field requires immersion in a specialized, complex, and jargon-filled world: a world of machines and their tenders. The primary challenge for untrained outsiders who examine workers is to learn enough to understand without learning too much, so that one can perceive the world with distant eyes. I explain relevant meteorological terms to assist readers without expertise in meteorology, as terms like vorticity, bow echoes, vil, zulu time, and mesoscale need clarification, even as I avoid still more technical terms. The American Meteorological Society–sponsored *Glossary of Meteorology* (2d ed.) weighs in at 855 pages, with 45 terms beginning with "z" alone. Some of the discourse is specialized, but informal talk also is common, such as "boxology," "knobology," a "dead pattern," or "flip-floppiness." My favorite was when forecasters told me that there was "no weather," meaning, of course, no likelihood of severe weather, disappointing for those who like the challenge of forecasting. Specialized acronyms like ETA, AWIPS, CWA, CAPE, AFD, NCEP, or DAPM are hardly transparent, even when spelled out.[8] At various meetings I found the range of acronyms and other specialized terms daunting and sometimes felt like an anxious pre-med who had spent too long partying. Acronyms, characteristic of the sciences,[9] are so evident that employees of the National Oceanic and Atmospheric Administration joke that the acronym NOAA stands for the National Organization for the Advancement of Acronyms (or, sometimes, for No Organization At All) or imagine the fantasy Bureau of Atmospheric Research Forecasters or BARF (Field notes). Sociology has a deserved reputation for jargon,[10] but, having been a sociologist for decades, I view *our* terms as transparent as air itself.

As in all occupations, metaphors abound in meteorology, transforming technical, scientific concepts into profane talk: clouds as sugar cones, pinwheels, castles, and beaver tails. Storms are popcorn, zappers, meatballs, and slop.[11] Radar images can be likened to a toilet bowl, Pac-Man, a beast, a bow, a battle,[12] or the Spanish Armada (Field notes). Weather naturalizes the social,[13] as society colonizes the meteorological. Physical features are given agency and motivation, and, thus, temperatures struggle to rise, rain wants to freeze, and storms shoot their wad (Field notes). In this, we see *nature's autonomy,* the belief that the sky has a mind of its own. We easily slide from claims of causation to claims of intention. Meteorological talk is a mix of images ripped from American culture and the local occupational subculture: talk that the public can understand and comment upon and talk that requires specialized training.

I was fortunate in finding patient teachers. Several forecasters, both older hands, the lead (or senior) forecasters, and younger journeymen (or general) forecasters assisted me on my travels toward meteorological literacy.[14] I recall one occasion in which I sat with Stan for his shift, and he patiently and carefully explained that high pressure was usually associated with "good" weather with the exception of so-called "dirty highs." He pointed to the synoptic (hemispheric) scale weather patterns, pointing to the ridges and troughs that tended to keep weather patterns stable over the winter months, explaining why regions often have extended periods of higher or lower than normal weather. Further, he indicated the pressure gradients (pictured as lines on a map of the height of air pressure), explaining that "the tighter the bars, the more weather." Being a careless consumer of two-dimensional weather maps, I had not realized how important that third dimension of height was in determining weather conditions: weather is not simply temperature and rainfall marching across the nation from Montana to Maine. Air, I was told, is like water, and the sky is "like an ocean" because of its waves and its properties of fluid dynamics (Field notes). The metaphor seemed useful to these workers and eventually to me. The key rule in meteorology is that "warm air rises and cold air sinks," emphasizing the three dimensional domain of the air. Further, as I was told many times, the critical fourth dimension of *time* affects weather systems.

On several occasions I was walked through the process of making a forecast, shown the information that the meteorologist examines in shaping a decision: imagining a funnel from the global meteorological community to the single forecaster on the hot seat. This information included forecast models, radar pictures, data from upper air balloons or aircraft, satellite images, and ground readings, as well as the forecast that had been distributed twelve hours previously. The amount of information was staggering, but, as in all such cases, those threatened with cognitive overload—such as air traffic controllers or medical students[15]—rely on communal heuristics, including pattern recognition skills, to determine which of these data are relevant. Forecasters have personal preferences, offices establish norms, and different weather conditions, including seasonal effects, give especial weight to particular types of data. Information is not created equally in the process of forecasting.

I began this research in January 2001 and continued until June 2002 with the exception of a convention of the National Weather Association that I attended in October 2002. The research was largely focused on the

Chicago (Romeoville, Illinois) office of the National Weather Service,[16] where I spent January 2001 until April 2002, present for three days a week for the first six months and approximately once a week for the remainder of the research. Since this time various procedures and rules have changed, but my account is based on that moment of weather service history.

The Chicago office is one of the oldest offices in the National Weather Service, dating to 1871. It once had forecast responsibility for much of the Midwest, but reorganizations that created a network of state and local offices removed this authority, leaving few more responsibilities than most other offices. But this distinguished history is important in understanding the emphasis on autonomy found in the office culture, an emphasis that has remained constant over decades. The office, once situated on the campus of the University of Chicago, is now located in a small suburb near Joliet, Illinois, but the main measurement point is situated at O'Hare Airport. Because it is assumed that most weather systems move from west to east, and often from south to north, many weather service offices are located to the southwest of large cities, as is the case in Chicago.

The Chicago office currently has the forecasting and warning responsibility for 23 counties in and around the Chicagoland metropolitan area, including Rockford and five counties in Indiana. During most of my observation, the Chicago office had a staff of twenty-four, including ten operational forecasters, an intern, three "hydrometeorological technicians" or HMTs (in charge of distributing the radio broadcast and gathering data), a science and operations officer (the SOO, pronounced Sue), a data acquisition program manager (the DAPM, "dap'm"), a public information officer or warning coordination meteorologist (the WCM, alternatively "wick'm" or "W-C-M"), a technology officer, two technicians, a port officer, a hydrologist, the Meteorologist-in-Charge (the MIC, "M-I-C"), and the administrative assistant (once the secretary). At the Chicago office, the administrative assistant, like most in the service, was female, as was the intern and the port officer. The others, including all the forecasters, were male. The two technicians and one of the lead forecasters were African American. The Chicago office is fairly representative of the National Weather Service in being dominated by white males, particularly in the higher status positions. The NWS does emphasize diversity, rhetorically and, as best I could tell, sincerely, but the combination of the job market, the stresses of shift work, and the image of who constitutes a proper meteorologist contribute to hiring patterns.

The other two offices at which I observed were similar: one had two female forecasters and the other had a Hispanic male forecaster, but diversity was limited in those (less urban) settings. Each of these offices had approximately the same number of employees in approximately the same job classifications. The Belvedere office was located in a rural area between the two moderately large cities that constituted their CWA (or County Warning Area). Belvedere is an established office and, like the Chicago office, once had responsibility for its state until the NWS created a network of offices with more compact areas of responsibility. The Belvedere office was established in the 1970s at the time that the National Weather Service decided to organize forecast offices by state. Forecasts for its area had once been issued by the Chicago office, but over three decades, Belvedere and other state offices became treated as older offices, even if they lacked all of the sense of tradition (and entitlement) of the Chicago office.

The Flowerland office, in contrast, was a spin-up office, established in the mid-1990s, part of the move toward decentralization. In contrast to Chicago and Belvedere, most of the lead forecasters were relatively young. The office did not have the same level of tradition, or as strong a culture, as the two other offices. I observed at the Belvedere and Flowerland offices ten days each, getting to know their routines, comparing and contrasting their practices and culture with that in Chicago.

National Weather Service local offices operate 24/7/365. In normal circumstances, at least two forecasters will be on duty in the office, although occasionally only a single meteorologist was present. Of course, no matter how energetic an observer, I could not hope to be a constant presence. However, in Chicago I observed two midnight shifts and at various points in the research observed every hour of the day and night. I came to be a fixture at the Chicago office, invited to parties and other outings. This was not as true in the other offices, where I was welcomed and instructed, but remained marginal.

In addition to these three offices, I also visited two National Weather Service centers. In order to explore decision making under pressure I spent two weeks at the Storm Prediction Center in Norman, Oklahoma, that shares space with the National Severe Storms Laboratory, a research center that is not part of the National Weather Service. Both organizations are affiliated with the prestigious meteorology department at the University of Oklahoma.[17] Like the local offices, the SPC is a small organization with a staff of approximately three dozen, mostly males. At the time I visited there were no female forecasters and one African American.

Like local offices, the Storm Prediction Center is a 24/7/365 operation, but with the mandate of putting out watches for severe weather, primarily for tornadoes and severe thunderstorms.[18] The SPC is particularly active from spring to fall with less activity in the winter months.

While the SPC ostensibly covers the entire continental United States, most tornadoes occur in the Midwest ("Tornado Alley") and throughout the Southeast.

While the tracks of large-scale weather events, such as hurricanes and major snowstorms, can be predicted with some measure of confidence, the advent of smaller-scale events, like tornadoes, are more problematic.[19] It was only in 1948 that the first warning was issued for a tornado that miraculously hit Tinker Air Force Base outside of Oklahoma City, Oklahoma, just as predicted.[20] For many years government forecasters were forbidden to use the word "tornado," for fear that it might create panic, placing it in a euphemistic netherworld with such linguistically problematic terms as "cancer" and "pregnancy." The first severe weather unit was established in 1952 and became the Severe Local Storms Warning Service the following year. In 1997, the office moved to Norman, Oklahoma, changing its name to the Storm Prediction Center.[21]

During my research I learned that it was not only observed data that allowed forecasters to make predictions, but "models." These models are extensive and elaborated sets of equations, based on theoretical assumptions about the nature of weather systems. They constitute the theory through which data are made available for forecasters to create predictions. Because weather systems are vastly complex and because meteorological theories are imprecise, these models are imperfect estimators. Weather is a chaotic system that is unlikely to be perfectly modeled. Models are expert systems, created by those with the authority to know, but with inexact information.[22] The importance of models to operational forecasting led me to spend five days at the Environmental Modelling Center (EMC) in Camp Springs, Maryland, and at the Hydrometeological Prediction Center in the same building, the center that forecasts significant precipitation.

I conducted face-to-face taped interviews at each of these locations. These interviews lasted from one to three hours and were bolstered with extensive discussions throughout the year. I interviewed eighteen employees at the Chicago office, including all the forecasters and meteorological technicians. This was supplemented by two interviews at Belvedere and two interviews at Flowerland. I also conducted three briefer interviews at the Storm Prediction Center and four at the Environmental Modeling Center.

I attended a four-day conference of the National Weather Association in Fort Worth, Texas. In contrast to the more academic American Meteorological Society, the NWA is a mix of academics, mostly those focused on descriptive meteorology, government forecasters, meteorologists working for private concerns, and broadcast meteorologists. The NWA split from the AMS in the mid-1970s, feeling that the AMS had little interest in issues of "operational meteorology." I also read several meteorology textbooks (beginning appropriately with *Weather for Dummies*[23]) and a large number of magazines and journals devoted to weather (notably *Weatherwise,* a bimonthly aimed at weather enthusiasts with a circulation of about 20,000, and the *Bulletin of the American Meteorological Society*), as well as more academic articles that touched on issues connected to the sociology of prediction or the history of meteorology.

As noted in the preface, I was welcomed in the meteorological community with considerable warmth. Weather forecasters, like those in many other occupations, feel underappreciated. Meteorologists, lawyers, cooks, butlers, butchers, police, politicians, and sanitation workers feel with some justice that their publics do not appreciate the conditions and constraints under which they labor, and this vexation provides an opening for ethnographers; observers are to be shown the "truth" of their plight. They wish us to describe their virtues and commitment. We do not always do as our informants wish, but this is their hope.

In this research I was informed: "You know the weather service is filled with weirdos," "you know we're all crazy," "it's a good thing you're not a psychologist," "have you found anyone who is certifiable?" "have you found out we're a bunch of sociopaths?" and a reference to "the madness of meteorology" (Field notes). Over the course of three decades, examining eight research sites, I have learned that one commonality among groups is the half-jocular claim of "insanity." I have playfully termed this *Fine's Law of Shared Madness.* These informants do not literally mean that they are mad but that they share subcultural eccentricities that those outside of their domain do not appreciate. Here, too, workers made the claim that the public assessments of their work and how this work should be done was at odds with their informal practices, necessary because of constraints and from a desire to have emotionally satisfying employment, a point emphasized by Everett Hughes, that pioneer in the sociology of work.[24]

With few exceptions I was accepted readily, not only in the attempt to teach me the business. One meteorologist asked me to complete a weather map for him, drawing lines between points of common pressure and similar dewpoints. Another permitted me to "play" on his computer when he wasn't busy. Once after I made a presentation to the Chicago chapter of American Meteorological Society, several forecasters brought in a cake with icing that read "Gary Just Fine." I was touched. I was also the subject of joking, both in my presence and absence. At times the forecasters claimed to arrange severe weather for my benefit. At one point they laughed that whenever I showed up, severe weather vanished—an outcome that caused mixed feelings as severe weather was emotionally enriching in contrast to mundane tasks.

Only occasionally were indications of (joking) suspicion evident, mostly at Belvedere and Flowerland, where, when I was taking notes, forecasters commented that "He's working for General Kelly [then head of the weather service]. Make him a double agent," or "Are you still spying on us?"[25] I suppose that there was some truth here, in that as a result of my presentation at the Chicago chapter of the American Meteorological Society, I was invited to present my thoughts at a Corporate Board meeting of the NWS in Washington. I accepted with some trepidation, but the invitation was made after my research was completed at the local offices and at the Storm Prediction Center, and during my research week at the modeling center. This lecture, and its outcome, with the talk shared with every science officer at each local office, proved helpful for establishing my credentials.

At various times I was tested, mostly in regard to shift work. Would I observe midnights? I did—twice—but these men had to do seven midnight shifts in sequence, an emotional and physiological strain. As one forecaster put it: "You have to work seven days on mids, and then we'll take you seriously. Unless you do, you're just passing through . . . Then we know you are worth keeping" (Field notes). He even attempted to entice me with the revelation that on their last night they have dancing girls and beer. However, his skepticism was just; I was only passing through. But they were kind enough to accept a tourist.

The History and Memory of Weather

Because of its impact on human activity, weather has long been part of human interest and social discourse. Animals of all species respond to weather conditions, and humans are no different. It is estimated that 90 percent of the American public check weather forecasts on a daily basis.[26]

Aristotle[27] is perhaps the first great weather analyst in his essay "Meteorologica," but he was surely not the first to speculate about the skies. By the eighteenth century, it was a common pastime for European and American amateur scientists (notably Thomas Jefferson) to inscribe weather conditions in diaries. Famed English naturalist Gilbert White kept detailed weather records for the village of Selborne for four decades.[28] Discussion abounded about the origins of meteorological phenomena, clouds among them.[29]

However, it soon became evident that the study of weather is not something an individual, no matter how conscientious, could do in isolation. Although meteorology is an observational science (only rarely experimental), it is a networked science. Much of the essential work, particularly in predicting weather, requires a network of observers. During the nineteenth century, proposals were advanced to map the movement of weather systems systematically. In the United States this mapping occurred under the auspices of the newly established Smithsonian Institution.[30]

Eventually, in 1870, the federal government created a national reporting system under the Army Signal Corps, an organization with observers already spread across the nation,[31] recognizing the linkage of technology with the military. Weather forecasting subsequently moved from the Department of War to the Department of Agriculture and more recently to the Department of Commerce, nicely indicating the changing importance of forecasting to various institutional spheres—defense, agriculture, and industry. The National Weather Service is one of the most visible government agency and one of the most respected, used by President Bush as a model of how government agencies should work. Currently the National Weather Service (formerly the U.S. Weather Bureau) is a part of the National Oceanic and Atmospheric Administration (NOAA).[32] Revealing continuity with its earliest institutional formulation, the NWS head is a retired Air Force general as was his predecessor (the head of NOAA during this period was an admiral). These leaders are felt to have superior management and organizational skills, but they are always called "General" or "Admiral," underlining their military connections. A considerable number of meteorologists had once been forecasters in the armed services.

The National Weather Service employs approximately 4,800 workers, including 2,400 meteorologists, located in 122 local offices, 13 river forecast centers and 9 national centers, including the Tropical (Hurricane) Prediction Center in Miami, the Storm Prediction Center in Norman, Oklahoma, and the Aviation Weather Center in Kansas City. The 2004 operating budget of the agency was $824 million. Over the course of a year, local offices issue approximately 50,000 weather warnings.[33]

The organization changed dramatically in the mid-1990s. Before that time, each state had a forecast office with several smaller offices to gather information and to warn of local severe weather. In the change the weather service established a network of offices serving comparable geographical areas, organized by county (the CWA or County Warning Area). As a result, the number of offices issuing forecasts more than doubled, leading to considerable hiring. These new offices were termed "spin-up" or "start-up" offices, and the older offices, such as Chicago, resentful at being "spun-down," often assumed that these forecasters were not as experienced (surely true) and less competent (more doubtful). While changes in technology, including the role of computers for data analysis and word processing, decreased the size of the organization, the new offices provided for additional staff.

This constitutes a brief version of the institutional history of weather forecasting. As Diane Vaughan emphasizes in her analysis of the National Aeronautics and Space Administration, *The Challenger Launch Decision,* organizations exist *in* history, embedded in institutional environments, and they exist *as* history, products of accumulated experiences.[34] I was impressed when forecasters recalled particular meteorological happenings, referring to them casually by the date that they occurred.[35] Weather records become important in typifying events, but the events themselves are significant as pegs for collective memory:

> Don notes that today is March 27, and he comments to Vic: "Eleven years ago we were issuing tornado warnings. It was just about now [3:30 p.m.] that a tornado warning was going out for southern Cook County, Lamont." (Field notes)
>
> : : :
>
> I was observing on a February day in which a sharp drop in temperature occurred (from 55 degrees to 38 degrees). Don started talking about April 6, 1972, telling us that the temperature dropped from 80 degrees to 19 degrees over thirty-six hours with five inches of snow. Don adds, "My wife would say, 'though he remembers those things, but he forgets his car keys.' I say I used up all my memory. There's little left." (Field notes)
>
> : : :
>
> At the SPC a forecaster tells me on April 27, 2002, "The Tulsa tornado was like this. Very late to clear out, and then F4." Another adds, "We still think about April 8, 1999, May 4, 1999. This could be a day like that. This is a big day. We've got four or five supercells just where we want them." The first forecaster says that there may be tornado watches until 7:00 a.m. the next

> morning, and he mentions a killer tornado in Cincinnati at 5:30 a.m. in 1999. (Field notes)

The importance of such recall was emphasized by one forecaster who recounted a tornado from fifty years previous, comparing it to the current situation. His point was that "no one saw [the earlier one] coming," and he wanted to make sure that this didn't happen again.[36] These are instances of occupational collective memory, a form of socially shared cognition,[37] common to all work communities. They belong both to individuals and to groups. Together the *office remembers* a wide range of events, and these events serve as benchmarks for judging current events by what had occurred previously. These memories justify the occupational claims of intuitive knowledge, separating these workers from those for whom weather is but a passing fancy.

Even though it flies in the face of formal scientific analysis, there is a strong belief in the importance of pattern recognition, particularly among older forecasters.[38] The emphasis on pattern recognition transforms meteorology into something akin to art, a personalistic and elusive process of interpretation, a domain of authenticity that is beyond the abilities or even understanding of outsiders. As geographer Mark Monmonier recounts in his incisive *Air Apparent:* "A storm's origins and subsequent movement could reveal its destination, but because weather is quirky, the forecaster remained alert for a sudden acceleration or shift in direction. This strategy worked much of the time, and what he learned from one storm, he filed away in mind and map to help predict others."[39] Attempts at *weather typing*,[40] creating a catalog of storm patterns, have been less than totally successful in that storms do not recreate their predecessors. The importance of microclimates and the chaos embedded in the system militates against rule-based knowledge. Thus, an informal system grounded on experience, incorporating the nuances of the atmosphere, is thought to improve forecasting over formal processes. Moreover, the emphasis on experience and pattern recognition provides forecasters with autonomy, carved from the domain of machines and models. This creates control over uncertainty, while simultaneously providing older forecasters with interpersonal authority. For example, one forecaster in Chicago contrasted the experience in his office with the absence of knowledge he saw in the newer, spin-up offices: "You don't have people who have been there 25, 30 years, who have seen everything. You have a lot of people who have been out for three to five years. You need more rules. We've had the experience in forecasting." "Seeing everything" is critical to meteorological authority. Said another, "That's why [the Flowerland office] is so wacky. They don't have the experience." (Field notes). As Neil Stuart, a workshop speaker at the National Weather Association meeting, asserted: "Use your own knowledge of past events. That's an important part of forecasting. That's part of pattern recognition" (Field notes). Another forecaster suggested that traditional career mobility in the National Weather Service requiring one to move from office to office to be promoted was counterproductive in preventing forecasters from gaining experience-based authentic knowledge of local weather.

These claims of intuitive knowledge, linked to group interaction, belong not only to the domain of meteorology, but to much scientific work and to other occupations that claim specialized knowledge. Memory and experience are central to the process by which scientific predictions are created. Both the organization and individual workers are embedded in history. Organizational forms—here the collective responsibility for forecasting and the creation of new spin-up offices—can either contribute to collective memory or retard it.

The Plan of the Book

As one focuses on any solid subject, it rapidly dissolves into a mosaic. Topics overlay each other like fish scales. Even with a focus—how small-group life affects work practices, occupational identity, and organizational culture—issues emerge that expand one's focus.

To select a single theme, not perfectly descriptive, I claim to examine the *production of the future*. By this I emphasize that the special domain of the weather *forecaster* is to *predict* what will happen in the future, establishing and drawing on the past, and then to communicate with a public so that others can act on these predictions. Forecasters focus not on the present or the past, but claim that they see the future. This future can only be known because of the establishment of what has gone before. The production of the future depends on the production of the past and present. In this weather forecasters share a prospective orientation with physicians, stock market analysts, political pollsters, tipsters, policymakers, commodity brokers, currency traders, economists, military strategists, acquisitions editors, fortune tellers, and parents. They think of themselves not as men and women of the here and now, but of the later, and this shapes how they interpret what they do and who they are, even if the later depends on what has been made of the here and now. In the case of weather forecasters, later may be a few minutes or days away (a close future), but it is a future on which people rely. Each chapter, but especially the third, addresses the problem of how to announce what is not here yet in light of what is and has been happening.

I set the stage by examining the practices of meteorological life. Chapter 1, "On the Floor," starts from the analysis of worklife to describe how forecasts are produced in real time. I examine the constraints of life in a meteorological office, both its routines and the emotional and temporal intensity when severe weather threatens. Because of the possibility of destructive weather, meteorology is a 24/7/365 occupation, placing enormous strain on workers, while investing them with authority. Their work entails keeping records of the past and communicating present conditions, but this is transformed into advice for the future.

The second chapter, "A Cult of 'Science,'" explores the *routine accomplishments* of these college graduates. Meteorology is classified as a "science," but what does this mean? It is, for the most part, not an experimental science, even in university settings, but it is more than descriptive in its prospective orientation. Meteorology is theoretical in that the creation of models is a professional goal. Yet, the practice involves generating predictions on the basis of current theory.

Science is at the heart of occupational identity within this government bureaucracy. Is the work scientific, not in an objective sense, but in terms of identity management and professional ideology? The concept of *scientist* is an honorific, and there is a push to claim that mantle. Yet, simultaneously these workers have culturally derived ideas about what science consists of,[41] and many are uncertain that this is what they do. Further, they are members of a discipline in which there are collective professional stakes, including authority, resources, and jobs. As a result, forecasters are ambivalent about whether they are scientists. The attitudes to science are also revealed in the ambivalence toward technology and equipment. Does the material inscription of the world provide autonomy or remove it? If machines have agency, as some claim, how do workers adjust to this reality? Can technology provide the authenticity that is central to the creation of a modern self?

In addition to belonging to an occupation, these workers are also members of small groups that like all groups develop a microculture or idioculture.[42] Offices are social systems with robust cultures—as experiments in the cultural organization of scientific practice and culture. They reflect the fact that all structure and culture are tied to local conditions: a sociology of localism.

In the third chapter, "Futurework," I explore the production of the future. How do meteorologists create forecasts to contain uncertainty? Meteorologists rely on gathered data in conjunction with models that provide a theoretical infrastructure. This affects the data to be collected. But if this was all that was necessary, forecasters would not be needed, so humans carve out a domain of personal expertise, selecting among alternate models, doubting the adequacy of data, and then adding their own experience. Armed with data, theory and experience, the organization provides legitimacy that is crucial for the presentation of these public predictions. Meteorologists, like other *future workers,* are authorized to predict by their sponsors. They are mandated to colonize the future.

Occupational autonomy, a frequent theme of those who examine work, is tied to organizational and group dynamics. Meteorologists wish to feel that they are constructing the forecast as authored. During my research, a period of intense technological change in the creation of forecasts, much of their autonomy involved control over language. I title chapter 4 "Writing on the Winds." Writing involves professional impression management and serves as the point of contact between forecasters and their publics. While I was observing, local forecasters faced major changes in work practices, threatening their identity, or so they felt. Headquarters was instituting a computerized forecast system, a system that largely removed the authority for writing a forecast from meteorologists. This change, allegedly providing workers with more time for meteorological analysis, caused great concern. Who has the authority to determine how weather is to be communicated? How is autonomy structured? How does this affect forms of coordination among forecasters and offices? I examine the literary battles of meteorologists, both as they were played out in the National Weather Service and as they were played out in particular offices as the culture of writing was being negotiated.

The fifth chapter, "Ground Truth," analyzes the organizational problem of scientific truth, tied again to the dynamics of group culture. The creation of models of truth has been emphasized in science studies, notably in Steven Shapin's *A Social History of Truth.*[43] But Shapin's analysis is fundamentally normative, examining the rules of truth-telling as set by society. In contrast, I treat verification as an interactional achievement. Forecasters make claims about what the future will bring, but these claims need to be accepted by others.

On what basis do we claim accuracy, how is this linked to our claims of authentic knowledge, and what are the practical issues in making these assessments? How correct must a forecast be? What is the line between right and wrong?[44] If a forecaster claims a 30 percent chance of rain, is the claim correct if it rains or if it doesn't? Does a high temperature of 52 justify the forecast, if the predicted high was 54? 60? 70?

Does a prediction of five inches of snow count, if eight inches fall? Verification is central to organizational control and the personal identity of meteorologists, but it is also a cultural phenomenon. As a government agency, the NWS feels a powerful push toward accountability. It must develop techniques that allow workers to evaluate how well they are doing, but these techniques are collective choices. Weather forecasts are responsive to group concerns; they are social constructions. This does not mean that forecasts are random—hardly so—but that they are tied to how particular communities define competence.

The sixth chapter, "A Public Science," addresses the reality that meteorologists have multiple audiences to satisfy and from which they want esteem. I use the label *public science* to refer to domains of scientific practice, like meteorology, that operate within the public sphere and with a primarily public audience. As members of the meteorologist's public, even if their assessments are translated and tweaked by the media, we rely upon their forecasts, and complain bitterly or drily when we feel that they have got it wrong. That these workers are tied to an organization, embedded in a governmental structure, affects prediction. Further, the system of media, commerce, and public interest suggests that each audience has expectations and demands. Private sources do not like government meteorologists to push too far into their domains, diverting profits and undercutting their claims of expertise. Government forecasters want to communicate to their public, but often this communication is mediated and shaped by others with their own economic and professional agendas.

The final chapter, "Weather Wise," addresses broad theoretical concerns and points to connections between this ethnographic investigation and other studies of scientific and work processes. I discuss core concepts of work, science, prediction, autonomy, truth, and public knowledge, hoping to generalize from this particular and peculiar case to other domains of science, of culture, and of work.

1 On the Floor

A sticky Chicago afternoon in August can feel like Hell.[1]

When I had last visited, the weather was cool and calm, edging toward autumn. As it was still summer, several workers who normally worked K shifts were on vacation. K shifts are day shifts in which forecasters work on their other responsibilities—updating the webpage, fixing minor computer glitches, compiling data for record-keeping, or planting flowers. A quiet lassitude defined the day, and the work seemed easy. Only two forecasters were on duty, one preparing the public forecast, and the other the Aviation forecast for O'Hare and smaller airports in the Chicago-Rockford area. They were supported by a meteorological technician overseeing the weather radio and compiling reports from a network of observers. The forecast was for more pleasant, mild summer weather, consistent with previous predictions that severe weather was unlikely. Forecasters distributed reports on schedule. The computer models fit the newly received data. Staff could glance at the newspaper, check favorite websites, sip coffee, and chat about family. The work done by three people could have been completed by one. The office was overstaffed. The administrators—the Meteorologist-in-Charge, the Science and Operations Officer, and the Warning Coordination Meteorologist—remained in their offices working on projects or completing paperwork.

Today the scene is dramatically different. Same office, different emotions. While no certainty exists as to what will develop, signs pointed to a day with "a lot of weather." As I describe below, a weather service office is an *activated organization*, often operating tranquilly but capable, like an emergency room, of becoming transformed into a hub of activity.

The Storm Prediction Center in Norman, Oklahoma, had issued a forecast that showed Chicago and most of its County Warning Area (23 counties) under a "moderate" risk for severe weather. Of the three categories of risk—slight, moderate, and high—moderate means that severe weather in the area is likely. The SPC rarely issues a high threat assessment, no more than a few times each year, and only when they believe that major severe weather is almost certain.

The Chicago office plans for severe weather to arrive in the late afternoon or early evening. Predicting whether tornadoes might hit is hard, but there are almost certain to be thunderstorms—boomers—probably with at least 3/4-inch hail and/or 58 mph winds, the criteria for severe weather. The atmosphere is highly unstable, the moisture level is elevated, a possibility of circular winds exists, and Kansas and Missouri had experienced severe weather late last night. The ingredients are present for *weather!*

If tornadoes or severe thunderstorm are about to emerge (or have just done so), forecasters must issue warnings. These warnings are for counties, and typically are within a larger area for which the Storm Prediction Center had previously issued a watch. The local office must determine that danger or threat from these storms is immanent before issuing a warning. Several warnings may be issued for the same storm as it travels through the forecast area. Severe weather can appear suddenly. This doesn't mean that storms are unpredictable or that weather systems appear "out of thin air," but the extent and intensity of storms may be unexpected.

The threat is such that the staff held a planning meeting in the late morning, discussing staffing and office responsibilities. The MIC informs the day shift that they will likely stay late, and a few forecasters who are not scheduled to work are called to alert them that they might be needed. The office is not sufficiently staffed for all of the extra work that the few days of severe weather bring, and staffing adjustments are necessary, as in hospital emergency rooms after a disaster.

Although George, the Meteorologist-in-Charge, is formally in charge of the office, Don, a lead forecaster with twenty years experience, runs the meeting, assisted by Byron, the science officer. Don and Byron describe the responsibilities of each staff member, including the administrators, the HMTs (the hydrometeorological technicians), and the forecasters. Don divides the warning area, permitting one forecaster to focus on storms entering the western part of the region, and another to concentrate on those in the southern and eastern counties. Storms typically travel from the southwest, and the airflow suggests that this will happen today. One staff member is assigned to gather storm information from ham radio operators, another is placed in charge of updating "nowcasts" and warnings, a third answers the phone, and a fourth updates the weather radio. The office is fortunate that a college meteorology student is working as a summer intern. He calls police and emergency workers in counties in which warnings had been or were about to be issued to learn what was happening, gaining a measure of what meteorologists refer to as "ground truth."

As a gesture of solidarity, George orders pizza, giving the day a festive quality. Since storms typically form in the late afternoon heat, in early afternoon there is little to do, except reminisce about past storms and desultorily monitor radar scans, satellite images, and data from upper-air balloons released near Springfield, Illinois, and Davenport, Iowa. The severe weather remains outside Chicago's County Warning Area. Don phones the Storm Prediction Center to gain their insight. He also calls forecasters in offices to their south and west, particularly colleagues in Lincoln, Illinois, and Davenport, Iowa. These offices are spin-up offices, opened in the mid-1990s, and Chicago staffers believe that they overforecast severe weather, but today all the offices agree. Marty, the public forecaster, distributes his seven-day forecast early to insure that this routine assignment will not interfere with his responsibilities for the incoming severe weather.

By 3:15 storms have entered the forecast area. Until they cross the border, the bad weather is not Chicago's problem. As an organizational matter, the thunderstorms do not exist until they near the county boundary, a political marker. These storms are strong, but the forecasters do not feel that they have reached the criteria for issuing a severe weather warning.

As storms move, tension, coupled with eager excitement, increases noticeably. These men care about the weather and severe weather tests their abilities. If a storm system is powerful, forecasters may become agitated, angry, or astonished, depending on their ability to cope.[2] As the heat of the afternoon builds, some storm cells increase in intensity. Vic, the forecaster in charge of the western area, worries about a storm cell southeast of Rockford. Don and Vic examine the radar images and decide to wait one more radar cycle (about five minutes) to see if the storm has intensified. With that new image, Don jokes, ostensibly making light of this important decision, "What the Hell! Let's do it," and they send out a warning for McHenry County, covering Chicago's northwest suburbs. Because of media demands and because of the organizational structure of emergency and police units, warnings are linked to county units. Science bows to political organization, just as when flu risk or unemployment data are presented on a state-by-state basis. Through the computer network, the police, fire departments, emergency workers, and media instantly receive the warning, also available through weather radio and on the office webpage. Richie, the college intern, phones police in the county to inquire about weather conditions, and, in the process, verifies the warning. The staff wouldn't be happy to learn that severe weather hit *before* the warning—that would have meant that they had missed the storm. Even a minute or two can make a difference between getting it right or missing it, affecting the office's success rate, a matter of concern for headquarters.

During the afternoon and evening storms pop up throughout the metropolitan region, popcorn storms as they are called. A funnel cloud is spotted on the beach off Afton, leading to discussion of whether it was a tornado or a waterspout, a matter of a few hundred feet, depending on whether it hit the beach or the waves. Since the office had predicted a tornado for Lake County, and since the person who called it in thought it might have hit the beach, they label it a tornado, grateful that they could claim that it hit land. Radar images themselves don't determine whether a tornado has touched down until forecasters receive a call or investigate it themselves.

The major event of the evening—the emotional centerpiece—is a tornado that hits two counties of rural Indiana, located within their warning area. The forecasters gather around the radar screen, gazing respectfully at a "hook echo," a classic indication of a tornado, because of its indication of strong circular wind rotation. They are impressed by the beauty of the colorful image on the radar, even while realizing the destruction it may cause. Although the Afton tornado is an F0, the lowest level storm, and apparently does no substantial damage, the Merrillville tornado hits a barn and destroys trees and roofs, possibly an F2 but more probably a strong F1 (on the scale from F0 to F5), a destructive tornado. In a storm this powerful, unlucky citizens may die, but not tonight. The next day, Patrick, the Warning Coordination Meteorologist, visits the area to assess the damage.

By early evening, heavy storms have moved from Chicago's responsibility; they have become the problem of the North Webster, Indiana, and Grand Rapids, Michigan, offices. The storms continue, but life in Chicago is easy. The extra workers leave by 9:00 p.m., some on duty for over twelve hours. Reports are prepared, calls are made, and there is nothing to do but wait until tomorrow, when storms may reignite, or not.

Working the Weather

The 2002 *Jobs Rated Almanac* rated meteorologist the seventh best job in the United States. The *Almanac* examined stress, physical demands, job security, salary, and work environment.[3] Sharing the top ten is financial planner; cowboys made the bottom ten—so much for preadolescent dreams. We live in a world of clean hands and indoor work, characteristics of contemporary meteorology.

In this chapter I explore the social contours of the meteorological life. My goal is not to be descriptive, although the description that I present may help those outside of the field gain a sense for the work. I wish to situate the occupation within its organizational and social psychological constraints. As in my study of restaurant kitchens,[4] I argue that the structure, culture, and interactions of operational meteorologists create the conditions in which weather forecasts are produced. In this case, it is the relationship of meteorology to science, to claims about the future, and to the communication of this knowledge that are at issue. I begin with the place and space in which meteorology is done, moving inward to work relations, the links between humans and machines, and labor under conditions of stress and threat.

Although meteorologists work for numerous organizations—universities, energy companies, utilities, airlines, consulting firms, and television stations—I focus on government employees, the authors of official weather information and keepers of the equipment that produces this information. As noted, the National Weather Service maintains 122 local offices throughout the United States to forecast for regions of a size that can be covered by the latest generation of radar equipment, about 60 miles in radius. Many National Weather Service offices were newly built or refurbished in the 1990s and are pleasant, if interchangeable. The offices I observed are modest low-slung one-story buildings, unrecognizable by their facade, but by the large radar dish behind the building.

Many are located on the west or southwest outskirts of the major city in the forecast area under the assumption that weather moves from west to east in temperate zones.[5] This location provides the radar the ability to detect incoming weather before it hits the metropolitan area. The Chicago office is situated in a distant suburb, some thirty miles southwest of the Loop. This location may subtly diminish the recognition that they are forecasting for an urban area, increasing the difficulty for them to feel the effects of their work on its end users. One MIC explicitly said that he did not want his staff to consider the impact of a warning, believing that warnings should be meteorological, not social.[6] Being in the middle of nowhere contributes to downplaying social impact and emphasizes that nature has autonomy.

The buildings are constructed with an eye toward severe weather; each has a small tornado room in the center of the structure in the rare event of a direct hit. The offices maintain a backup electrical generator, switched on when severe weather threatens. As a result, forecasting can continue even if the weather is dangerous. The structure of the building is linked to assumptions of the structure of weather[7] and affects meteorological thinking. The architecture and location focus attention on severe weather. As in many workplaces, the tasks of workers are made concrete in the design and placement of their spaces.

A casual visitor would not realize that the building is other than an office. There had once been a suggestion—a seemingly odd one—that the buildings should not have *windows*, preventing the weather from distracting the meteorologists from their technical assessments based on readouts from machines. This suggestion was shared with me several times, intended to indicate the foolishness of government bureaucrats. The plan was never implemented, but these claims were expressed to justify the authenticity of lived experience over mechanical readings. On this occasion observational experience triumphed over technological assessments. Forecasters treasure their windows and assert that they help in forecasting. One explained, "I thought it was nice that my laboratory was right outside the door." Another said, "They wanted us to use the equipment [instead of viewing the sky]. That's our backup—to look out the window" (Field notes). They are "in the field" at the same time as they are examining machines. Windows are valued at moments of severe weather; when skies are clear, they are irrelevant, and forecasters may pull the shades to enable them to see their monitors better.

The building is divided into a large central area and a set of small offices or cubicles, a kitchen, and a conference room. The ten forecasters lack private offices, sharing their workspace. This structure suggests a critical difference between public science and academic science. There is no place of retreat; the individual worker is not treated as an autonomous creator of knowledge but as an organizational cog, part of a corporate enterprise with only the local culture to provide some measure of autonomy. As David Livingstone[8] points out, science work is channeled by its locations. The structure of weather service offices emphasizes that these workers are producing collective products, not linked to personal perspectives, leaving only the writing of the forecast as a matter on which a forecaster can place his stamp (discussed in chap. 4). The idiosyncratic flavor of scientific creation is wrung out of these government forecasters.

The forecasters work in the central area—referred to as *the floor*[9]—the size of a large living room, perhaps 20 by 30 feet, filled with several dozen computer monitors, keyboards, and CPUs, work areas, and cubicles.[10] There is also radar and radio equipment, and, depending on the office, technology for measuring upper-air weather. This is, in effect, their lab, their shopfloor.[11] One section of the floor is set aside, in Chicago, for the *public forecaster*, the meteorologist who sends out the weather predictions for media and public use. This person was described as being *in the hot seat*, particularly on days with *weather*. The chairs—in effect, the office of the forecaster—are comfortable, padded, swiveling, leather seats. It is *in* the chair, as much as *at* the computer, that the forecaster does his or her work. Another section of the floor is reserved for the *aviation forecaster*,[12] who creates the weather forecasts for local airports, notably O'Hare, but is also responsible for the forecasts of weather conditions over Lake Michigan.[13] A third area of the floor is set aside for the hydrometeorological technician. These workers are in charge of distributing information through the weather radio and the Internet, gathering weather data (observations, or "obs"), including checking outside instruments every six hours, and insuring that contact is made with the cooperative observers, men and women spread out over the region who collect climate data for the office. Some of those who do HMT work are interns, able to bid on forecaster positions after sufficient training, while others have a career track that keeps them as HMTs. Many of these latter workers lack college degrees in meteorology, and many were trained by the military.

A striking difference between these offices and academic science offices is the absence of personal decorations. Individuals do not own space but borrow it. There are no places for pictures of family, diplomas, or personal knickknacks. One staff member evaded this restriction by opening a notebook with photos of his wife and children when on duty.

The bulletin board was open for all to post items of communal interest and amusement, but the question of who was responsible was never resolved. Some items remained for weeks; others for days.

Spatial control is central to organizational life.[14] Space is a resource and a source of power. The question of who "owns" a space is real, and sometimes resource battles are fought over a cubicle or a bulletin board. During my observation, the forecaster responsible for the office webpage actively resisted his space being "taken" from him for general use. He claimed that the construction of the webpage was of sufficient importance and involved specialized programs that his cubicle and its computer should be dedicated to his use. The question was never fully answered, although he continued to have priority in the cubicle.

More often issues of control involved the transitory use of space, although this fleeting use didn't make struggles less significant. As Christena Nippert-Eng emphasizes, the temporary use of space constitutes a ritualized transition,[15] both between workers and in creating a boundary between "pre-work" and work itself. At the Chicago office, Windex and Lysol helped to demarcate temporal boundaries and were sprayed shortly after the previous user left the building. Although not every forecaster did so, and some used Windex for computer screens and plastic desktops and others the more symbolic Lysol, the message was that a transition had occurred.[16] The space became newly minted. The cleansers constituted a moral boundary reflecting an attempt at control. The cleansing rid the space of pollution, creating "my" space from "your" space.[17] While one could make a case for the need for these sprays for aesthetic or hygienic reasons, they have symbolic value. Thus, when Don arrives for his evening shift, he sprays Lysol on his work area. Byron comments, "You want to make sure that everything is sterile," and Don responds, "I'm worried that nobody else is." (Field notes). That this control function is recognized by workers can be seen when Randy, in replacing Sid, sprays his space with Lysol before Sid has left, and Sid jokes, "What do you think I am, the blob?" (Field notes). The norm requires that the previous occupant has left the building; otherwise the erasure of pollution is too obvious. Forecasters joke about which colleagues need their workspaces cleansed, emphasizing its moral and evaluative dimension.

Less significant morally, but indicating control over the shift to work, is the desire of forecasters to change the color combinations on their monitors when they arrive, making the weather theirs. Again a communal resource becomes symbolically personal. One forecaster was known for what others considered garishly bright indicators of weather lines.

None of the color schemes, of course, represented what was happening in the air; they were aesthetic indicators of numeric data, but the choices were assigned meaning within the group culture. Individual forecasters favored particular models or different types of data (e.g., pressure levels at particular atmospheric heights). Meteorologists need not rely on the same information in creating forecasts and can access their own favorite sources of data. While forecasts are collective in that they build on each other, symbolic choices personalize the work.

As the space *belongs* to everyone, changes, even relatively insignificant ones, can gain political meaning. The local union became involved in the placement of a printer at one office, and on another occasion the height of the partitions separating desks was at issue. These arrangements are assigned meaning in a collective, negotiated place in which occupational autonomy is important and is reflected in the smallest issues of the right to control space.

In the Hot Seat

Routines are integral to work lives and identities, providing both order and constraint. Even when routines are disrupted, they remain the basis on which the normative order of work is judged. The existence of routine makes a job seem doable and provides a measure of control over uncertainty. In conditions of strain or crisis, workers know that they will eventually return to a routine.

Yet, routines are never absolute, but adjustable. Work is not a clichéd image of an assembly line; even an assembly line isn't. The emotional tenor of the office and pace of work varies. As in restaurants, forecasters' work fluctuates by day and by hour, even if, unlike restaurants, the "rush" is hard to predict much in advance. Still, late afternoon to early evening, as the atmosphere heats up, is most likely to produce severe weather and organizational stress. Forecasters may forget during their routine work that they, like police and firefighters, can save lives, but when severe weather threatens, this thought—powerful, frightening, and empowering—intrudes. As one explained: "People don't realize what an important job this is to save lives. . . . You have to analyze storms with lightning speed. This is a critical job. You have to be on an on-call basis" (Field notes). Yet, the frequency of calm weather—seven of ten days have no precipitation in Chicago—may make the job cushy, leading to secondary involvements. As one groused about his colleagues, "I was raised in the weather service where you're there to work. I don't want cable or writing your girlfriend or day-trading or sleep" (Field notes).

While I observed web surfing, errands, and others personal activities, they are no worse than the involvements of university faculty.[18] The organizational question is whether one is paid for working set hours or working to complete a task. If the latter, then these other activities barely matter, except to skeptical outsiders and insiders with the thorny eyes of strangers.

Most of my observation occurred during the day shift. In Chicago workers arrived at 8:00 a.m. for the eight-hour day shift (7:00 a.m. until 3:00 p.m. if they were on the Aviation desk). Most days begin quietly. On the first of a series of day shifts, the forecaster examines the past few days to "get up to speed." Otherwise, the forecaster examines the forecast from the previous night (the 4:00 a.m. forecast), the current model guidance, and new measurements as that information arrives. Little forecasting occurred before 11:00 a.m., and sometimes not until after lunch. By this time the data from the upper-air balloons released in the early morning are available and models are updated. In Chicago the forecast package was supposed to be distributed by 3:30 p.m., although the timing varies slightly from office to office, in part a function of media pressure. The workday changed with the introduction of the computerized Interactive Forecast Preparation System (IFPS), a technological innovation that altered how forecasting was done,[19] but until that time workers felt that on quiet days, forecasts could be easily compiled in two hours. Sometimes in the morning forecasters work on other projects, and occasionally they would run an errand or purchase lunch with others covering should the unexpected happen. Staff claimed that on most days, a single forecaster could complete the work of two.

Meteorologists base their predictions on previous forecasts and the guidance from a set of models distributed by the Environmental Modeling Center. When the models are fundamentally in agreement, the task is easy. As Joan Fujimura rightly notes, science is a form of collective action.[20] Even though one person creates a forecast, he does not do so independently, but depends on information compiled by others.

The process of developing a forecast is known as *knobology* or *clickology*, borrowing the scientific ending *-ology*, transforming their technical interface into science itself. The terms, however, are awash in irony, seemingly scientific, yet implying that the job of the forecaster is *merely* to twist knobs and push buttons. In bowing simultaneously to science and to mindless labor, it symbolically addresses the striving for scientific authenticity in a bureaucratic domain.

Forecasters amass data from various domains, a process of *articulation work*, defined by Adele Clarke and Joan Fujimura as "the invisible and unacknowledged but often arduous work of pulling various elements together in the 'right' sequences and at the 'right' times and places in order to achieve particular goals."[21] Emphasizing the importance of work routines, each forecaster can describe his own work practices, practices that often follow a ritual cup of coffee.[22] I present a lengthy, but abbreviated account from one forecaster to show the kinds of information gathering and assessment that is needed for competent work on a "typical" day:

> First you come in and get a briefing from the midnight guy, and every shift I start by shutting down the work station and bringing it back up to clean out all the processes that build up [on the computer from the previous forecaster's shift]. And then I like to load specific procedures that I have. I usually have one graphics monitor set up for current weather, obs, satellite, radar, and the other things that are happening. And I'll keep the other monitor basically for looking at all my model stuff. Mostly I'll look initially at what's happening currently. I'll look at satellite, look at what the temperatures are. Is there anything moving? And maybe I'll have a model up just so I can kind of get an overview of what the next few days are going to look like. Even though I've gotten a briefing, the next thing I'll usually do is I'll actually read the AFD [area forecast discussion composed by the midnight shift], just so I fully understand what his thinking was and see what we're up against. And then sometimes, if I've got time, I'll sit down and actually do a hand and surface analysis just to assess where any frontal boundaries might be nearby or look for something that's going to impact us right away.

Thus, the first task is to bring oneself up to speed, gaining a sense of the recent past and how that past might be extrapolated into the future. From here, the forecaster waits as new data arrives during the shift, suggesting possible futures:

> You're waiting usually about three hours before the new model data is really coming in. That is usually between 9:00 and 10:00 Central Time here when that stuff starts showing up. And then the Aviation model usually comes in an hour and a half or so behind that. So really the rest of the morning is spent just cruising the model data as it comes in and whatever else comes up. If a zone update needs to be done [because of changes in weather

> conditions], obviously that takes precedence, and that's again where I like to see what's actually happening, so I know what's going on and then monitor that as the hours progress, and to make sure our forecast isn't getting out of hand. You monitor that as you're looking at some of the new stuff.

Finally about noon the forecaster moves from reading the weather to creating it: first in his mind and then in written form:

> [Noon is] when I really start making my decision for the zone package, so usually by that time I've chosen my model or group of models or what I want to do. So I'll look at the different models, and I'll look for glaring errors. And what I'll do then is I would usually sit down and take some notes. I'll actually have a pad of paper there, and I'll go through day by day, and I'll take some notes for myself for creating the AFD, and for keeping things straight in my mind, because when you're looking at so many different time periods and different models, it's easy to forget where I was, to lose my train of thought, and then I also use those notes as I'm actually preparing the zones and to keep my timing straight. [At this point the final data arrive]. Probably between 1:00 and 2:00, [I create the extended forecast]. In the real far out periods, I'm not really looking to do a lot of detail. I'm looking for trends, which day might have a shot at precip and general temperature trends. And once I have that decided, and I've got all my notes, I'll go ahead and write my AFD. Pretty much once that's done, it's time to be typing the actual zone forecast. (Interview)

Even though forecasters are interesting to watch, accounts of their routines are less compelling. Still, this account emphasizes the merging of numerous information streams, including models, observations, satellite images, and radar. The weather is not there to be forecast, but data must be gathered and processed through standards of competent work, choosing those sources with which individual forecasters are most comfortable, creating an authentic forecast, based on personal insight. Meteorologists are challenged to tame the range of physical indicators and model predictions, producing a plausible claim as to how the future will unfold.

Each forecaster has preferences. Despite the controlling force of routine and institutional demands, some personal choices remain. With the individual forecaster and the local group attempting to "own" their distinctive products and the organization attempting to establish uniformity, the tension over to whom these predictions belong is real (see chap. 4). This attempt to insert self in meteorology creates an authenticity of scientific claims, even though the end products are roughly comparable, suitable for coordination with those before and after, and roughly consistent with predictions in surrounding areas. Each forecast depends on the collective access to data, what Clarke and Fujimura describe as a theory-methods package.[23]

Forecasters control the timing of work, provided they adjust to when information is available and to when the organization insists that the translated information be distributed: the input and output boundaries of a public science. The organization of a workday is responsive to the needs of media and publics. The afternoon forecast provides information for drive-time radio and the evening news. The early morning forecast provides predictions for morning radio and television. The weather service coordinates its products with the temporal requirements of influential corporate users. This creates temporal patterns, and, on occasion, strain. In the hour prior to the forecast, the tempo of the office quickens,[24] producing the rush or what one forecaster refers to as "crunch time." In this, forecasters are no different from other workers—radiologists, bus drivers, disc jockeys, or butlers—in which the demand for services at particular moments is set by others.

Forecasting and the Division of Labor

Organizations create webs of tasks and expectations that permit assignments to be completed efficiently and predictably. A classic method for achieving collective ends is to establish a division of labor, an allocation of responsibilities, often linked to segmentation of time and space. These positions carry differential status and authority. Although scientific work may be thought of as communal, as famously suggested by Robert Merton,[25] science is also a status hierarchy.[26]

Administration and Forecasts. The central administration of the National Weather Service has established several administrative positions to direct and provide support to the forecasters and meteorological technicians. Beyond the meteorologist-in-charge and the administrative assistant (or office manager, formerly the secretary), the National Weather Service assigns each office a Science and Operations Officer (SOO) and a Warning Coordination Meteorologist (WCM). The former has the task of accessing scientific meteorological knowledge and representing that knowledge to office staff; the latter connects with the public and severe weather spotters, becoming the public face of the office. Each office hires a hydrologist, who forecasts flooding potential, a task beyond the expertise of many meteorologists. Offices also have a specialist in information technology and a staff member in charge of their equipment, the Electronic Systems Analyst. Coastal locations hire a Port Meteorological Officer, who in Chicago communicates with boat owners and ship officers on Lake Michigan. These positions reflect the institutional priorities of the National Weather Service, a set of linkages to influential publics.

Focal Points. Just as a division of labor is created by establishing different positions, a division of labor exists within forecasting. Most MICs perceive a need for specialization and establish a set of assignments for forecasters that are labeled "focal points," referring to areas of responsibility. Among the focal points in Chicago are winter weather, short-term forecasts, aviation, webpage development, building and grounds, severe weather, fire weather, radar, weather radio, climate, and lake forecasts. A staffer does not do all of the work in an assigned area, but is responsible for that area. The focal point is expected to develop expertise. Hiring decisions are tied to replacing the skill set of a staff member who has left, never leaving critical focal points open. This recognizes the collective nature of the organization, creating a system of distributed knowledge.[27] The sum of knowledge is not known by each individual but is available through enforced specialization. Forecasters rely on each other. As one forecaster explained, "Once you get the title [as a focal point], they will come to you. Because you have the title, they think you're God. They expect you to know everything about the system. By having the title, they expect you to have the expertise" (Interview).

Forecasters and Techs. A third division of labor is more closely linked to the status system in science, a division between scientists and technicians.[28] In the case of operational meteorology this involves the boundary between the forecasters and the hydrometeorological technicians. This latter group is divided into two categories: interns, who with the passage of time (generally two or three years, if they "pay their dues" and pass a series of tests) become forecasters, and met techs, for whom the position is permanent. Under the current system of training, interns have B.S. degrees in meteorology and occasionally some training in the private sector or in the military. In the past, forecasters could transfer from the military without the credential of a college degree, but today the degree validates a theoretical knowledge base that affirms forecasting as a *scientific* occupation. The HMTs do not have that option; it is a dead-end career path, which traditionally didn't require a B.S. degree, but, often, military training with a minimum of two years of college.

The proximity of these interdependent status groups sometimes results in occupational tension, as in laboratories, hospitals, and law offices.[29] The HMTs, less likely to see themselves as integrated professionals, are more likely to support the employees' union, frustrating forecasters, who value their decision-making autonomy. In return, forecasters occasionally denigrated HMTs, at least in the eyes of these workers:

> With the change in shifts, eliminating the midnight HMT shift, forecasters had to compile a report that had previously been done by HMTs. A forecaster complains that now he will have to send out "the HMT report." He describes the book that details how to compile the report, "It's like HMT for Dummies." He adds in front of Martha, an HMT, "If they want to overpay me, I don't mind." He clearly sees this work as beneath him. Another forecaster comments, when other forecasters leave for the night, "We'll stay with the low price help." (Field notes)
>
> : : :
>
> Mitchell, an HMT, comments, "I don't have an analytical mind. [But being in charge of the co-operative reporting program] You have to work with people. You're dealing with people skills." He adds about being an HMT, "This side is considered lower class. Second-class citizen." Another HMT comments about one of the more science-oriented employees, "I'm the lowest form of life in his eyes. Totally." This HMT notes about a call from a forecaster at another office, "He wanted a forecaster to call, not a peon like me." (Field notes)

These observations are bolstered by comments, sometimes quite tart, from HMTs:

> As a whole the office gets along fairly decently, but I think a lot of times that some of the forecasters think of us as glorified secretaries. They look at us to answer the phone calls and do things like that. Like a lot of times to write a Nowcast [an

> update of current weather conditions] they don't trust us to be intelligent enough to do that. (Interview)
>
> : : :
>
> The opportunity [for HMTs] is not there because of the way the weather service is set up. They built a barrier with a sheepskin, and they said, "If you don't have this sheepskin that says the right words, you can't do this job." Fine, I can play that role.... [The forecasters] will be talking in a situation, and we'll mention something that is pertinent to that kind of stuff or at least something that they probably need to listen to, and it's just like we weren't there sometimes. And someone else will say almost exactly the same thing, but with a different sheepskin and [he snaps his fingers], the same words. (Interview)

These comments could easily be duplicated, as with the forecaster who remarked of HMTs that "they are not really part of the fraternity" or the stories of the retired forecaster who, it was claimed, refused to have HMTs talk to him. They may be treated as non-persons. While the status barriers are eroding, in part due to friendship and in part because forecasters increasingly are forced to do the work of the technicians, the barriers are real. The job of the forecaster has been to determine what the future will be, the job of the HMT, focused on collecting observations, is to determine what the present is and what the past has been. In this, there is some similarity with the doctor from whom we expect a prognosis, and the nurse who takes real-time observations of temperature and blood pressure and the outlines of a medical history. HMTs are support personnel for the creation of forecasts. Even if this is functional for the organization, it still erects a status boundary.

Partners. The fourth system of distributed expertise is based on the fact that forecasters work in pairs. Offices arrange schedules differently, affecting the strength of relations between partners. In the Belvedere office partners were "married to each other"—one lead forecaster routinely worked with one journeyman forecaster, modified by vacation schedules. A more experienced man[30] was paired with a less experienced one. The older man was supposed to teach and supervise the younger one. Even though at this office, they shifted between the higher-status long-term and lower-status short-term forecast position, a status system based on seniority was built into the organization. When the partners worked well together, things went smoothly. One of the Belvedere "couples" carpooled together. One senior partner commented, "Dan always knows where I'm going [what he is about to do]." The two continually joke with each other, and they compare forecasts to insure that they are consistent. He adds, "we can tell from our banter, who is doing what." They claim to have developed intuitive knowledge of the other, making two persons one.

In contrast, the Flowerland office created a schedule in which forecasters are not routinely paired, although typically for each shift a lead and a journeyman will be on duty. This is much less collaborative than the Belvedere system. Rather than the image of "being married," they refer to the problem of the "odd couple." Perhaps because they are a spin-up office Flowerland staff assume that everyone has something to learn from everyone else; this way younger forecasters can benefit from the knowledge of each senior forecaster. Training is broader but is less intense. One lead forecaster noted that "In severe weather, it's really easy for me to staff up, because I know everyone's strengths and weaknesses" (Field notes). For him this system works for both the older and the younger forecasters. These differing perspectives were articulated by the MICs of Belvedere and Flowerland, respectively:

> [The role of partners is] very important for this office. Extremely important. . . . They learn to work with each other very well, and they tend to augment each other with their talents. If one doesn't have this talent, the other one does. I know one forecaster's good at this and another forecaster isn't good at that, so [in] that weather situation one forecaster takes the lead, and he may be a good radar operator. He'll jump on the radar rather than the other forecaster. . . . I think there are more benefits of having the strong teams than disadvantages. I see the people are happier on a day-to-day basis because there are people that they are comfortable with. (Interview)

: : :

> The cycle that we have for rotations for senior forecasters and the rotational cycle for the journeymen and the interns are completely different in that there's a constant mix so that you work with everybody. . . . There are always going to be times if you work with the same person all the time, sooner or later one of those individuals is going to take a vacation, and they're going to have to be replaced with somebody else. If you're so ingrained in working with the same individual all the time, you get into

> the habit of what to expect from the other individual, and if you bring somebody else into it who might be accustomed to working with a totally different person, you have a different work habit something may fall through the cracks. Try to mix things up and you learn more. Different people have different habits. You might take a plus or minus. If somebody does something, "Hey, I like that. I'll try to remember and do that next time." You can learn from it. Chip Henry is a very good trainer for being able to recognize events and how well the numerical weather model handled certain events. If only one person worked with him all the time, that would be the only person that might get the benefit of his knowledge. By rotating it through, everybody gets the benefits of what he knows. (Interview)

These quotations indicate that offices have the organizational authority to create their own microcultures and that each of these local structures can be justified. The former permits an implicit and taken-for-granted community, while the latter, with weaker relations, demands a more direct style of speech.[31]

The Chicago office is something of an admixture of the two systems. Each forecaster has three partners: a partner for the day, evening, and midnight shifts. The lead forecasters rotate backward (day, midnight, evening), and the journeymen rotate forward (day, evening, midnight). The success of this system, particularly for the more secluded evening and midnight shifts, depends on the character of the partners. Some are compatible; others work side-by-side without sharing meteorological knowledge: one team of two or two teams of one.

These partnerships are also status systems with a senior and junior partner, even though, as in so many domains, status may be discretely hidden. As one lead forecaster explained about a period twenty years previous, "I used to think of the lead forecaster as a grand poobah." Under the older system, nothing was distributed to the public without the lead forecaster's approval. Today one cannot tell status by work tasks, and some journeyman forecasters are older—or have spent more time in the Weather Service—than some lead forecasters. With the exception of a recently hired journeyman, I did not see a journeyman ask the lead to check a forecast. On only one occasion was this an issue. Here the journeyman asserted that the lead was much more likely to warn for severe weather than he was, and when asked about a storm cell by an intern, he explains with some asperity, "My professional opinion is that my lead forecaster will tell me to warn" (Field notes). This instance was striking in that junior forecasters are typically given considerable autonomy. They can request confirmation but rarely is the status hierarchy explicit.[32] Once you are in the club, all are equal publicly.

Shifts

Weather offices—like the weather—operate on a 24/7/365 schedule, an important constraint on organizational culture. They are more like public safety offices—police, military, fire, hospital emergency—than academic laboratories. In a society in which most individuals work "normal" business hours and in which temporal consistency is prized, the structure of shifts for weather forecasters is a problem. Part of the explanation is that "weather" can appear suddenly, but another part is that the self-image of the organization is tied to the military,[33] the agency that over a century ago had the responsibility for forecasts. While offices differ in how they arrange schedules, most forecasters are expected to work a "week of midnights" every five weeks. The typical structure of shifts involves a day shift from 8:00 a.m. to 4:00 p.m., an evening shift from 4:00 p.m. to midnight, and a midnight shift, lasting until 8:00 a.m.[34] That shifts are *rotated* is critical to the strains on the organization and its workers. One handbook on shift work[35] suggests that over 25 percent of the American workforce have "nontraditional work schedules." Among the problems pointed to are chronic sleep problems (60–80 percent), stomach disorders (4–5 times greater), chronic fatigue (80 percent), mood swings and depression (5–15 times more likely), higher rates of drug and alcohol abuse, more severe accidents, and higher rates of divorce and spouse abuse. It is a hard life. One forecaster estimated that he has given fifteen years of his life to his country, the time he imagined his life will be shortened by working shifts. *The Shift Workers' Handbook* hopes to solve these problems by seeing shift work as "much more than just a work schedule. It's a way of life. When our schedules change, our entire lives change." Shift workers must see their schedule not simply as a technical matter, but as both a biological concern and as a life-style.

In many organizations, workers work a shift for an extended period, permitting bodily adjustment. Factories may maintain a 24-hour schedule, but workers often work a set shift. Even the police, organized by shift, do not change shifts on a weekly basis. The structure of the weather service insures that all workers share equally in the temporal responsibilities of forecasting. Making this more problematic is that not every worker in the organization rotates shifts. The administrators work a normal day schedule, unless there is a need to fill in. The Belvedere office established a single ten-hour shift for HMTs (6:00 a.m.–4:00 p.m., four days a week), and the other offices have two eight-hour HMT shifts. Only the forecasters alter their schedule with costs to health and family relations, while creating a solidarity in which forecasters perceive themselves as a "band of brothers,"[36] an emotional intensity that overrides the disadvantages of the profession. The division was so evident that one administrator described himself and his colleagues as "day weenie folks" (Field notes).

Each shift has its own character, producing forms of work culture.[37] During the day shift, workers chat with colleagues and may work on other assignments. They operate within a normal work schedule,[38] even if the cultural and physiological effects of previous shifts have not dissipated.

The evening shift has a reputation of being undemanding, so much so that the Flowerland office sometimes assigned a single forecaster for duty during this shift. Meteorologists might update the previous forecasts, but otherwise had little to do. They were not responsible for a new set of forecasts. The evening shift was described as when you "babysit the forecast." One forecaster particularly disliked evening shifts because he felt that he was not of much service. He was not where the action is.

Midnight shifts have their own rhythm. Forecasters send out their forecast at 4:00 a.m., and this requires evaluation of current weather conditions, observations, and models. If the weather is stable, however, and no severe weather threatens, forecasters could "tweak" the afternoon forecast. The day forecast has priority, and day forecasters are more likely to change the midnight forecast than the reverse. On the midnight shift, only rarely were major changes made to the extended forecast. After issuing their forecast, some staff slept fitfully, others surfed the web, and still others did personal work. Only two people were on duty and as long as they reached an understanding of the occupational proprieties no one else would be the wiser. It is an unwritten rule, and one I did not see violated, that a forecaster was not supposed to take vacation during a week scheduled for midnight shifts. The problem of midnight shifts is such that the NWS labels a coffee pot as essential equipment, permitting the office to purchase it from the facilities budget.

The effects of such a system are multifold. First, it motivates career trajectories in that forecasters often accept administrative jobs to avoid shift work, just as cooks hope to become daytime executive chefs or cops hope for desk jobs. Others claim that they will retire as soon as they have reached their thirty years of government service, seeing no reason "to beat myself over the head with midnights" (Field notes). Still others attempt make-dos, adjusting to their schedule as best they can, and hope that their family can cope. Being married to a shift worker is not easy. Several forecasters recounted divorces and other marital strains.

Rotating shifts create a barrier for hiring and retaining female forecasters, creating a *gendered occupation*. In a world in which fewer forms of labor are now gendered, operational meteorology remains so, and the gendered organization of the occupation is seen as an essential and unchangeable feature of the work. I was told repeatedly—by men and women—about "the obligations that [women] feel toward [their] family" (Field notes). Rotating shifts justifies a masculine occupational domain; as one former military man asserted, "If you can't do shifts, you're not a man." Another spoke of "macho mids." A third comments, "To me it's insane. Why would someone sabotage their life. It's a pseudo-military thing" (Field notes). While it was constituted as "guy talk," the joke reported in the introduction of having "dancing girls and beer" on the final night of one's shift suggests that working mids is considered male time. This organization of time is gendered—symbolically and in practice. While this is surely not the only reason that of the thirty forecasters I observed, only two were female, the temporal structure of the organization, as it is transformed into symbolism, was a factor, given the expectations of and for young women.[39] Of course, at one time it was assumed that women would not tolerate full-time employment or continuous night shifts, so even this seeming allergy to rotating shifts may change with time. Women do not work rotating shifts because women are not *supposed* to work rotating shifts. The organization of labor constrained each aspect of the doing of work, posing a challenge to the establishment of a professional discourse, as it does for emergency workers who cannot set the terms and pace of their own labor.

Severe Weather and the Activated Organization

Ultimately weather forecasting is linked to public safety; we want our meteorologists to protect us. Knowing the temperature is nice; warning us of possible death and destruction is necessary, even if few find themselves in the maw of an approaching tornado. It is in the near-term prediction of the future that meteorologists earn their salary, and our praise and contempt. As the National Weather Service explains in its mission statement: "The National Weather Service provides weather, hydrologic, and climate forecasts and warnings for the United States, its territories, adjacent waters and ocean areas, for the protection of life and property and the enhancement of the national economy." This routinized organization can become *activated*, a matter of life and death, converting its temporal routines. It is in the oscillation of time that the core of occupation life in an organization that must deal with crisis can be discerned.

While rare in any place or on any occasion, the amount of severe weather can be daunting. In the 1990s an average of 876 Americans died annually from weather-related causes, with estimated losses of over $13 billion a year. Tornadoes kill 56 people each year on average. This was minor in comparison to extreme heat (282 deaths) and extreme cold (292 deaths). Tornado deaths were fewer than those from floods, winter storms, and lightning. The annual dollar loss (in 1999 dollars) from tornadoes was $777 million, less than floods, hurricanes, and hail.[40] But the number of reported tornadoes has been increasing, a function of better radar, public education, and more trained tornado spotters.[41] One assumes that the actual number of tornadoes in "God's own count" has not changed unless perhaps global warming plays a role. In the Chicago warning area from 1986 to 2001, there were 113 tornadoes, of which 65 were warned for (only 21 were predicted by tornado warnings, the others were predicted by severe thunderstorm warnings).[42] Most warnings for both tornadoes and for severe thunderstorm were not verified—at least for the county for which they were issued—in that no evidence of tornadic activity was reported. During this period only one violent tornado hit the Chicago area, the deadly F5 tornado that devastated Plainfield; only one other tornado was classified as much as an F2 on the six-point F-scale (F0 to F5). The F-scale (or Fujita scale) divides tornadoes into six categories, based on the damage they inflict, although the damage is supposed to indicate the circular wind speed, which, because of the evanescent quality of tornadoes, can rarely be measured. An F0 tornado, the vast majority of occurrences, has wind speeds of only 40–72 mph. Despite their fearful image, most tornadoes are little more than a strong wind, less than hurricane strength.

The National Weather Service has styled itself in what seemed a parody of an advertising slogan as "the no-surprise weather service," appropriate for a no-risk society. One recalls the old slogan for Holiday Inn that the "best surprise is no surprise," treating uniformity of service as a moral virtue and eliminating the personal, authentic touch. In practice, this means that forecasters wait for indicators of the unusual and unexpected to appear so that they can be tamed through institutional rules. As noted, the National Weather Service is, like a hospital trauma center or a fire department, an *activated organization*, routinely operating on a low-energy level, overstaffed, except on those occasions in which it is transformed, verging on being overwhelmed and understaffed, until routine can again be established.[43]

Although some argue that all weather is important, convincing forecasters of this is an uphill fight. Sunlight, for instance, leads to high rates of skin cancer, drought, or solar power availability, but measuring sunlight is said to be "boring" and "stupid," an idiosyncratic passion of bureaucrats at headquarters who have forgotten or never knew what it is like to be in the trenches.[44] Severe weather is the beating heart of forecasting. Sunlight is treated as "no weather," as local organizational imperatives and emotional responses shape the contours of what is occupationally significant. Sunlight does not require an activated organization.

To examine how the staff of the National Weather Service handles severe weather as an organizational concern, I first describe how workers cope with severe weather in developing an emotional structure of work, particularly coping with blown forecasts. I then explore the process by which forecasters "see" severe weather. Finally, I address the issue of timing of *watches* and *warnings*, particularly in light of the uncertainty of the interpretation of severe weather.

The Emotions of Prediction. Meteorologists root for severe weather. They are primed for uncertainty. In this, they are little different from physicians and attorneys who dream of tough cases. Crises test their professional mettle, even if their clients would prefer something less dramatic, finding some way to make the tumor, trial, or tornado just go away. It is not that these workers desire death and destruction, but they do hope for occupational memories,[45] for war stories.[46] They hope to be "where the action is."[47]

Records make for recall. One meteorologist told me that during the deadly 1995 heat wave in Chicago,[48] "we were all looking at the temperature, saying 'go up, go up'" (Field notes). They stood on the cusp of history. Even if they realized that they were unable to control the skies, the skies belonged to them as translators. They had the power as reporters of the future. A forecaster told me about the threat of severe weather: "It gets the adrenaline going.... It's like an overtime game when you're on the edge of your seat.... a [calm] day like today, you don't have the same level of anxiety." On another occasion, too mild for his liking, he mused, "let's see if we can get some winter" (Field notes).[49] Said another, "Tornadoes are beautiful. I don't want to see anybody hurt, but I want to see *action*" (Field notes).[50] When one suggested that a particular storm cell looked "wishy-washy," another joked, "We need more super-cells" (Field notes).

In the emotional universe of meteorology, the optimist was the man who felt that there would be severe weather; the pessimist was convinced that the weather would remain quiet. This was made explicit when I was told, "when we say good, we mean bad for everyone else. A good storm is a bad storm" (Field notes). On another occasion, one forecaster remarked, "That big outbreak today is in trouble," with his partner replying, "You see it not panning out. It's sometimes painful to watch. I guess the good news is . . . instability is getting better in the area" (Field notes). When a storm or cloud wall bears down on the office, the forecasters line up at the window, just as doctors share unusual X-rays. These shared, authentic experiences, establish membership in the group through a collective past, specialized knowledge, and claims to authority. This emotional entrainment[51] tests occupational skills and legitimates one's self-esteem. Emotional energy is crucial to occupations in which discovery and prediction are part of the mandate. The cultural centrality of this affect is dramatically expressed in a story I was told on several occasions at the Storm Prediction Center, partly to socialize me but partly because my informants felt that it linked to core issues of identity. As one forecaster recounted: "There was a TV crew in the office, and there was a deadly tornado, and the forecasters were giving each other high fives, because they had the watch out twenty minutes before. Everyone who was a meteorologist knew what they were doing, but they should have known better" (Field notes). Significantly the director told me this same story when we were planning my visit, perhaps in part to direct my attention away from what an outsider might interpret as a lust for lethal weather. To be socialized to severe weather involves being able to unpack this story and to understand that the issue was that they had the opportunity to use their skills to exert control over the uncontrollable and also the fact that by tasting the emotional juice of a deadly twister, they come alive.

It is common for scientists to identify with their objects of study. I don't know if research oncologists root for rare cystoblasts, but biologists identify with their objects of study, whether they be salmon[52] or fruit flies.[53] This affiliation sometimes expresses itself in dreamwork. Although not every meteorologist reports dreaming of the weather, some do. Randy shares a dream in which he is leaving the office and sees a half dozen tornadoes bearing down on the parking lot. Similarly Patrick recounts a dream in which he sees a white Cadillac flung through the air by a tornado (Field notes).

The emotions engendered by severe weather are not always happy ones. The uncertainty and temporal demands of forecasting add powerful stress, as Chandra Mukerji suggests for oceanographers.[54] These forecasters are competitive, and dread blown forecasts, particularly if they miss a major storm. One meteorologist put it, "I have a passionate desire to win, and so we bleed inside a lot." Blown forecasts are inevitable, given the uncertainty of the air, but that doesn't make them easier to accept.

Blown Forecasts

> The sorrow of mistakes is sometime very diffuse and sometimes very pointed. It is sometimes the sorrow of failed action and sometimes the sorrow of failed conduct. The sorrow of mistakes has been expressed as *the too-lateness of human understanding* as it lies along the continuum of time, and as a wish that it might have been different both then and now.[55]

Workers must cope with the inevitability of errors, and then account for their cause and distribute blame.[56] Institutional structures, individual delicts, and "chance" are said to be responsible. Of course, some occupations, notably those that engage in futurework, are particularly likely to be linked to error, and as any consumer of weather information knows, operational meteorologists are not known for their inerrancy.

One who desires certainty and can't abide error should not choose meteorology as a career, no more than they should be doctors, financial planners, or fortune-tellers. As one forecaster sighed, "It's not the type of job to get into if you're a person who always has to be right. You won't last long." Another noted, "You can't have much of an ego for this profession." A third joked, "The worst we'll be is wrong. . . . I figure if it happens, it happens." One forecaster who had worked for the Quantitative Precipitation Forecast office in Washington, charged with forecasting all significant rain events in the continental United States, told me that they were correct only about 35–40 percent, and for events with 2 or more inches of rain they were correct only 10–15 percent, leading a colleague to wonder if they could have forecast Noah's flood. As forecasters joke, "Never let current weather affect your forecast," to which they add, "never let reality affect your forecast" (Field notes).

Because of the range of competing alternatives one meteorologist explained that "the [lengthiest] forecast is the one we know the least about" (Field notes). In practice, when a meteorologist misses a forecast he may apologize to his replacement who had to mop up afterwards. One forecaster decided not to put thunder and lightning in his forecast, and the next day he consoled (and blamed) his replacement, "Sorry about last night. I told you to update the forecast. A short wave came around" (Field notes). Another forecaster had predicted gale force lake winds that did not appear. He commented to his replacement, "Things aren't coming along the way I thought they were going to do. Maybe gale isn't the way to go. I'm thinking about backing off. . . . That's what amendments are for" (Field notes). A third occasion involved a prediction of sunny skies in which the forecaster attempted to apportion blame. He apologized to his replacement for the blown forecast, explaining to me, "The model data was putting us into sunny skies. I tried to put clouds into the optimism of the forecast. My temperature forecast was a complete bust. I was off by ten to fifteen degrees. It was incredibly bad. I made a partial adjustment, but I should have known better. I apologized, 'that forecast was really a dog,' but I really got hooked into guidance. He was a party to it. I had inherited his forecast. He was a party to the busted forecast" (Field notes).

Apologies usually receive a sympathetic hearing, even while they distribute blame to preserve one's identity as a competent professional.[57] At times forecasters insert a chance of rain or snow to protect their claims. Announcing a 20 percent chance of precipitation avoids embarrassment in the case of rain, even if small likelihoods of precipitation are erased in media reports. Further, since meteorology is a collective enterprise, dependent on information received through guidance, observations, and previous forecasts, blame is more easily apportioned.

As in so many occupations, errors provide for socialization, although what is learned may be narrow, perhaps avoiding that error by overreacting. I was told, "If [forecasters] miss one, they will overwarn for the next one or two" (Field notes). A forecaster at the Storm Prediction Center explained that "if an office has a killer tornado event, they will warn a tornado for everything" (Field notes).

Some errors are severe, tied to a loss of life and severe property damage. How do forecasters cope with the reality that their decisions may have profound consequences? Uncertainty is hard enough, but certainty of error smudges, or ravages, the self. A blown forecast is linked to self-esteem.[58] Some defend themselves with sarcasm ("No matter what you say, you're going to be wrong"), while others respond by emotional detachment. As Jake comments: "Some forecasters take [errors] home with them, but they get used to it. If your forecast goes bust, you think not 'I blew this,' but that 'the forecast got blown.' Otherwise you'll never make it" (Field notes). The passive voice preserves the self. Explained another, "You're going to fail out here. There's no way you can be right every time. It's hard psychologically. When a major storm occurs, you always ask yourself could I have done anything differently. Mentally you have to keep yourself positive" (Field notes). Jake recalls the time when a colleague on the Aviation desk sent out a forecast that contributed to a plane crash, "You got yourself a huge blown situation. . . . You always think, 'what would happen if my forecast was bad, and you're responsible'" (Field notes).[59]

The issue is not simply the emotion per se but that it occurs within a context in which the future did not develop as the forecaster expected and, as a result, discredited the organization. The forecaster, anonymous to the public, embodies the agency. One forecaster brooded that for a long time he needed a stiff drink after each shift because of one of these events. I was told of one forecaster who missed a deadly tornado, "Psychologically he was finished for the rest of his life. If you don't have a watch out, it affects your life" (Field notes). Some forecasters resign, become clinically depressed, or battle alcoholism.

Organizational imperatives contribute to mistakes, even deadly ones, but individuals have agency and their choices can lead to disaster. Rules are promulgated, formal and implicit, to refrain from mentioning others' blown forecasts,[60] symbolically keeping identity threats at bay (chaps. 2 and 5).

Keeping Cool. As in so many other occupations, portraying the proper emotions is believed to contribute to proper outcomes. The cardinal virtue of meteorologists confronting severe weather is the ability to "keep cool." They must become skilled in emotion management and the presentation of those emotions, embracing the regulation of affect. Forcing oneself to stay cool leads to a calm assessment of dangerous weather, a core feature of emotion work.[61] An administrator at the Storm Prediction Center indicated that the ability to remain calm—not to "yell and scream"—was a characteristic that he demanded in his lead forecasters. The frustration of error must be internalized, whether the error came from missing a cue or because one was misled by colleagues or machines. So much information is processed that any misinterpretation or neglect could be critical. Those who lack the ability to inspire confidence in others in coping with mistakes are terminated or not promoted. This again reflects the military traditions of the weather service. Knowledge of meteorological patterns was not the only grounds for success. He explains:

> There were definitely some [journeyman forecasters] whose meteorology was better than the lead position... but the position is more than just issuing the watches. It's leadership. It's professionalism. It's teamwork. It's coordinating with the office and staying cool under pressure.... We're only picking the people that have all the qualities that I mentioned to you. And if they don't have those qualities, they may say, "Hey, I'm the best meteorologist."... They'd have to change their style or there would be problems. We had an interesting couple of leads that are no longer here. One of them lost his cool several times on the desk.... He didn't come across as being confident.... You got some fear, you can't ever show it. The leader, you got to make sure that you stay calm.... You've seen someone who's kind of gone crazy, and I have seen stress get into people where they literally lose it. (Interview)

The MIC at Flowerland was adamant about stress management. He reported that one of the forecasters "had words" with another, and in a culture in which anger is devalued and seen as indicating illness,[62] this created organizational peril. The MIC explained that this forecaster had been working on a complex weather situation and the computer froze and it appeared that his hours of work were lost. He "snapped at" one of the administrative staff, with his face "beet red" and (allegedly) his blood pressure high. His frustration was defined as unacceptable emotional deviance. The MIC remarked that this forecaster could have been cited for insubordination and slapped with disciplinary action. In response to his concern (and in view of his conception of his role as father confessor, as opposed to a commanding officer), the MIC told his staff: "There is a lot of stress showing up with all the new products.... You're doing a great job. Weather comes up. Sometimes you're right. Sometimes you're wrong. Just do the best you can.... We've got to handle it better, so we don't become sick. Heaven forbid someone has a heart attack or a stroke because of the stress" (Field notes).

The most chilling sentence I heard was uttered by a research scientist with the National Severe Storms Laboratory, which shares a building with the Storm Prediction Center. Every day at lunch, the two groups met for a discussion of current weather patterns (about fifteen attended, more or less equally divided). In a discussion of a set of powerful tornadoes some years earlier, this scientist turned to an older and esteemed forecaster and joked, "How many people did you kill, Fred?" Everyone laughed. That he made this comment emphasized the linkage between the forecast and the outcome. Such a remark could have been intensely threatening, but that this comment was directed to one of the most esteemed forecasters meant that no one would take it seriously, while still making the point (Field notes). The scientist did not, of course, have the luxury of saving—or taking—life. His stakes were tied to theory.

If there is danger in error, the satisfaction in meteorology is in "getting it right," a theme that arose repeatedly. I asked a forecaster to describe his feelings of job satisfaction.

> The same feeling that a paramedic would get if they revived somebody, or that a policeman would get from saving somebody from harm's way. When we issue tornado warnings, . . . in the event that a large tornado does evolve from a particular storm, you have a warning out for it, and you hear on the news later that people had a five-minute lead time to take cover and that's why there were no deaths and maybe just a few injuries. I mean that is a wonderful feeling. That would be what we live for. For those moments or putting out blizzard warnings. Now people laugh and make jokes about the idea that it's almost like from the outside that people think that it may be a little bit sadistic. I mean we love it when the weather becomes threatening to life and property. It's not that we want to see death and destruction, it's that it's a challenge. That's what we're here for. I say like when a fire alarm rings in the fire station, the rush that they must feel. (Interview)

These workers live for bad weather so that in their predictions they can gain control and satisfaction.[63]

Accurately predicting the future receives flickering notice, occasionally and briefly. In contrast, errors are more salient—shoveling 6 inches of partly cloudy, as my father-in-law put it. A meteorologist remarked, "Because you get it right, no one pays it much attention. That's the kind of thing you have to deal with. If you're wrong, you have to swallow your pride. There's no good way to eat crow" (Field notes). One must accept the emotional blows that the skies deliver. Forecasters continually discuss being wrong, perhaps being self-effacing, but also providing a grounding for excuses if they err with false positives or false negatives.

Even after the fact, the forecaster (but not colleagues) can mention the mistake in symbolically significant ways. On one occasion, a forecaster brought in a pie as *humble pie.* A second case was more dramatic. After missing a forecast, in this case putting out an announcement of moderate risk for severe storms on a day that remained calm, an SPC forecaster entered the office wearing a grocery bag with eye holes cut out. On the bag was written "Chicken Little" (Field notes), referring to the hen[64] who feared that the sky was falling, a self-effacing prank that was referred to throughout my two-week visit. He judged himself not macho enough, too afraid of the possibility of severe weather. But equally important, he was part of a community of judgment and was convinced that his co-workers would see him in the same way, and so his display was an indicator of role-distance, suggesting that he saw his mistake as they did.

Forecasters must deal with the reality that weather events can be unpredictable. This unpredictability—*the agency of the skies*—constitutes a problem that prevents workers from preparing for a sure future but also provides a legitimating excuse. At times the range of possible short-term forecasts is broad. A forecaster comments about such a day: "If you have severe weather, you know what you have to do. . . . If there's no weather within 500 miles, you know what to do. This is the most annoying kind of weather. I'm in limbo. I want to know what I'll be doing" (Field notes). Another forecaster mused, "We just sit here and wait and watch. That's the job of meteorology. You just sit here and wait. You won't know the outcome for a couple of hours" (Field notes). The occupational challenge is tied to one's expectations, preparing to respond. While forecasters enjoy the challenge of severe weather, they dislike being surprised:

> Sean tells me after a day in which he had to forecast for thunderstorms, hail, and funnel clouds (none of which were severe): "A little too much action when you're not expecting it." (Field notes)
>
> : : :
>
> Sid discusses a day in which they were expecting tornadoes, but none developed, "That was just a pain in the ass. I don't mind being busy if there's something going on." (Field notes)

As with an occupational rush generally,[65] severe weather compresses temporal experience[66]: in the words of one forecaster, "if you have severe weather coming in, the night just flies by" (Field notes). Time has different organizational implications, affecting the social psychology of experience.

Seeing and Predicting. Ensconced in an office, sitting in a comfortable chair, staring at a row of monitors, how is one to know what is happening *out there*. What is the process by which tornadoes are seen? This is a scientific equivalent of Erving Goffman's[67] question of the framing of social interaction. Goffman asks, "What is it that is going on here?" What cues are we to use to make predictions of the future; what are the most salient variables? Observing rain, winds, even dead calm, are we to expect tornadoes, and how do we know? Tornadoes in the Upper Midwest have a character different from those larger, frequent, and more powerful twisters of the southern plains of Kansas, Oklahoma, and Texas. Chicagoans are unlikely to share Dorothy's experience, transported, dog in arms, to Oz. Plains tornadoes often develop from supercells, massive thunderstorms. Tornadoes in the Upper Midwest often arise from a line of thunderstorms and heavy showers, the leading edge of a front, often referred to as bow echoes from their shape like a hunting bow. This difference in storm size and formation affects how forecasters in the Upper Midwest think about the storms that might—or might not—appear.

In chapter 3, I return to the problem of forecasting for a future—several hours to ten days. Here I address the immediacy of knowing what is about to happen—sensing a wild world while remaining indoors, reading from machines and listening to human accounts. Machines create a form of representation that brings images of the outside world into the laboratory, so that the world is simplified and used for purposes of prediction. These images are not the things themselves but indicators of these things (for instance, the amount of reflection measured by a radar indicating moisture in the air). Instrumentation provides a transformation of the world in such a way as to make it usable for purposes of prediction.[68]

Reading Images. As opposed to other weather systems in which upper air data, models, or satellite pictures are helpful, for tornadoes the instrument that integrates humans and machines is the radar. The Doppler radar is the forecaster's window into unstable skies. The Doppler radar currently used by weather service offices (WSR-88D) is, like all machinery, limited by its engineering. The radar is a large spherical instrument topping a tower located at the weather service office. The radar generates short microwave pulses (about three billion/second), and a receiver picks up signals that bounce back when the waves hit objects, such as drops of precipitation. The waves recognize differences in air density and can measure whether objects are moving away or toward the radar, including rotating winds that signify possible tornadic activity. Meteorologists have developed indicators, most famously a "hook echo," a small curved area of intense wind, that warn of the possibility of a tornado, and a "descending core" that suggests the presence of hail. Each indicator is read from bright squares of color generated through radar scans. Some of these images are read visually and some the machine translates into other measures, such as the amount of "vil" (or vertically integrated liquid), indicating the extent of moisture, or the amount of "reflectivity," a measure of moisture reflected through the radar pulses. High reflectivity aloft suggests that the main updraft of the storm is holding potential hail aloft, giving the forecasters more time before severe weather reaches the surface (Field notes). Forecasters look for signs in the image of the storm cell (a "hook" or "ball") or movements (a "right-moving storm") that are taken as markers of severe weather, replacing inaccessible direct observation. Either the machine or the human can be the source of error, but it is humans—supervisors and forecasters—who assign blame.

The radar does not "picture" the sky but only reflects objects and their movement. These pulses are transmitted diagonally from the top of the radar tower. The direction of radar pulses takes into account the curvature of the earth. As a result, at various points of the radar's range different levels of the sky are depicted. The farther from the tower, the higher the points of the atmosphere the pulses measure. Near the tower the signal can be masked because of what is labeled *ground clutter*, too much interference when the pulses are close to earth. As a result, a radar image cannot be read as transparent; it must be interpreted. A storm at the edge of the range might be poorly sampled or be different in character from one closer. Storms too close to the radar may appear as part of the ground clutter and are only discovered if the clutter appears to move. Because radars have overlapping ranges and because signals are more or less sensitive (depending on whether forecasters are looking for severe weather or normal precipitation), some mechanical eccentricity can be managed, but radar images are not photographs, even if taken as such.

Given that it takes the radar five minutes to complete a circular turn (approximately ten minutes in "clear air" conditions), forecasters must decide whether to act on the current image or to wait a few, possibly crucial, minutes for the next scan. Is it better to be certain of danger, to avoid "crying wolf," or will those extra minutes of warning be crucial? Is it better to make a type I error or a type II?

The rhetoric of seeing in real time is commonplace. One forecaster explained of a possible tornado: "This is really rotating. The whole thing is curving. . . . You can actually see the rotation there. It's really impressive" (Field notes). He claims to see what he cannot see, except through a conventional representation. What is "seen" is only recognized because a machine has presented a conventionalized image.

When the data available do not "make sense," contrasting the radar data with the ground-level indicators or indicators from the upper-air data, forecasters create stories that explain the differences in ways that members of their occupational group can accept, or rule the contradiction irrelevant through joking ("maybe it's the traffic on the tollway"), maintaining their predictive control. In each case, negotiating or ignoring alternative realities, they create a workable image of meteorological reality.[69] This occurs in a situation that requires quick action or none at all. Severe weather—"convective weather," small-scale, but powerful—requires "a different mindset" from the slower, more extensive winter weather systems. The first severe weather event of the year is often a challenge, because, like drivers who face their first snow of winter, meteorologists must activate intuition rusty from disuse. As one forecaster put it: "It seems that we have an acute problem in the spring when convective season starts to get busy. We all need to get calibrated and after a long winter of not very many events, we're a little out of practice, and it's difficult to keep that sharp focus on what you need for severe weather when you haven't done it for three or four months" (Interview).

Hearing Reports. These judgments, based on mechanical read-outs, have their counterparts in human inputs. Forecasters do not only see, but they communicate. On days in which severe weather is considered possible, the office will be fully staffed; forecasters or administrators are called in as needed and those who are supposed to work on subsidiary tasks are brought onto the floor, the main forecast area. The organization, now activated, is a small group with overlapping responsibilities, sometimes uncertain, leading to the assessment that an office during severe weather is characterized by "controlled mayhem" or "controlled chaos." Consensus-building and negotiation are routine. Frequently someone is assigned to "play God" or "traffic cop," but as offices are based on collegiality, authority must be muffled through an *implicit deference structure*. Authority must be tacit and unstated, organized through indirect communication.

In addition, an office is set within an institutional communications network in which forecasters receive input from other institutional sources, notably other offices and the Storm Prediction Center. Nearby offices are the most immediate source of information and constitute their occupational peers, men and women facing the same meteorological and organizational challenges. Convective storms generally move from west to east, south to north, and so offices to the west and south "pass off" storms. Forecasters may call offices where they have friends to gain a history of the storm system with more ground truth than the radar provides. These forecasters have lived through the storm and can reveal its character. The offices in whose territory the storms are entering do the calling. It is impolite for the call to go west to east.

A second source of information is the Storm Prediction Center, the organization responsible for assessments of severe weather risk and watches. The SPC issues a risk assessment for three days ahead, updated on the second day, and again for the current day. As noted, the risk can be slight, moderate, or high. Local offices often call the SPC if there is a moderate or high risk for severe weather. One day the Chicago area was under high risk, but despite the organizational activation, nothing happened; the assessment was a bust. Still, several calls were made during the day to coordinate.

If the SPC believes that there is a likelihood of severe weather in a more narrowly defined area, they will issue a watch. Before they do, they call the local offices that will be affected, developing a consensus and establishing support for their decision. Although the SPC can override the wishes of a local office, they often accede and "watch boxes," parallelograms of risk,[70] may stop at the boundary of an office's responsibility. A watch has implications for office staffing, for the activation of emergency offices in affected counties, and in generating calls from the media. A watch increases a local office's workload and may frighten citizens. Since a watch does not indicate that severe weather is imminent, many offices ask for a firm indicator of danger. Discussion can be sharp, if politely gloved. As one Chicago forecaster explained, "I try to discourage them from putting out a box unless they have a good reason" (Field notes). While my informants in the Chicago office objected to being labeled "conservative" or unwilling to accept legitimate watches, their colleagues in Oklahoma felt that the Chicago office required more justification than other offices. Perhaps Chicago meteorologists were right in hesitating to frighten the residents of their large metropolitan area, but their caution posed an organizational problem for the Oklahomans, who felt their autonomy questioned and their decision-making slowed. Local workgroups are ultimately set within a communications network of individuals confronting similar problems, but with different perspectives or with different scope for action.

Timing Risk. Imagine yourself a meteorologist faced with the *possibility* of severe weather: how great a likelihood would you require before you issued a warning, saving—or disrupting—the lives of fellow citizens. Of course, there is no way of establishing precise likelihoods as we do not think as statisticians, but the question is valid. So valid, in fact, that when one MIC returned from a regional meeting he emphasized that he learned that offices had varied criteria for issuing warnings: "They asked what percentage [of likelihood of a tornado] would you put out a warning. I didn't know. I never really thought about it. I put down 50 percent. The range was from 10 percent to 80 percent. The guy at the 80 percent, you got to shake that guy pretty hard before he puts out a tornado warning" (Field notes). Should we prefer false positives or false negatives? To cry wolf or accept destructive surprises? Do you need to "see something" or are suitable conditions sufficient? As I discuss in chapter 2, local offices have different reputations, different histories, and different philosophies that produce these choices. Chicago is believed to be conservative, cautious about issuing warnings;[71] Flowerland, a new office, warns more often. There are many hypothetical assessments when severe weather is a possibility, as when forecasters discuss a storm system for which the SPC issued a watch that ended at the boundary of the Chicago area:

> DON: That squall line is starting to activate.
>
> SEAN: I have a sneaking suspicion that as this stuff crosses the [Mississippi] river, it will fall apart.
>
> PATRICK: I suspect when the sun goes down, it will disappear.
>
> SEAN: It looks healthy. I wouldn't be surprised for them to drop a box on us.
>
> BERT: If I were a betting man, I'd say it wouldn't [hit us]. (Field notes)

Although one might imagine that forecasters would rather put out warnings, even when they are uncertain, this is not always the case. There are costs to both false positives and false negatives, and forecasters and offices have different ideologies of risk. Risk assessment is not objective but derives from social, cultural, and institutional choices:[72] a "social transformation of risk."[73] Social and technical decisions cannot be separated,[74] as natural scientists are transformed into social scientists, perhaps against their will. If the National Weather Service chooses to label itself "the no surprise weather service," that affects their warning policy, pressuring forecasters to avoid missing a storm.

Those who wish to warn "early and often" give primacy to nature's autonomy: what could be worse than a tornado ripping through a residential neighborhood, hitting a school? Most drunk drivers cause no damage, most nail clippers in carry-on luggage are innocuous, and most storm cells are harmless, but at times we decide that a small risk of a terrible event is worth preventing, even if that prevention imposes a small and certain cost on many.[75] One often hears that "the risks of overwarning are less than the risks of underwarning" (Interview). As a meteorologist put it tartly, "It's better to say, 'Gee, I'm sorry it didn't work out. You didn't get whatever your forecast is.' Better that than to have to say, 'Gee, I'm sorry you lost your house or you lost someone in your family because you didn't get time to take cover'" (Interview). Yet, how can you measure the counterfactual of "lives saved." Staff narrate accounts of forecasting heroism, forgetting the wasted efforts:

> "Why did you needlessly alert us, when nothing happened?" I've heard that reaction from people.... It's almost like they were mad that they weren't hit by the tornado. "Why did you warn me? The tornado didn't hit me." You should be happy it didn't hit you. You would think that would be a good thing, but there are some people, they don't want to know unless it is a clear and present danger.... It cuts into your credibility. If you say too many times that something's gonna happen and it doesn't, it's the old 'cry wolf' syndrome.... I don't want to miss events either, because that's where people die.... There's what we call an asymmetric penalty function in this business, which means there seems to be a lot more to lose by missing events than there is to gain by minimizing false alarms.... If we miss a couple of important events, we get skewered in the media. Congress pulling funding.... The local offices, the natural response is that they overwarn for tornadoes. (Interview)

However, there are costs of overwarning, leading the public to be skeptical of warnings, wondering if the routinely claimed threats are real.

Missing severe weather means that the forecaster is accepting some forecasting risks in an uncertain environment. As one veteran told an intern, too ready to warn, "You want to put it out when there's really something going on" (Field notes). Too much lead time and the message can be discarded or made irrelevant by atmospheric changes—the future is always dynamic. Thus, forecasters worry about the decision to warn:

> This tends to be a conservative office, and that seems to have been passed down from generation to generation of forecasters here. And actually when it comes to forecasting, I think that's actually a good thing, because day in, day out, most of the time, the things that happen are relatively normal and within the bounds of expected conditions. And so to forecast outlying events—big snows, big floods, they don't happen that often and when they do happen usually all the things have to come together just right. . . . When I first came here . . . there was sort of an attitude, sort of a cautious, "I'll believe it when I see it" attitude. (Interview)

: : :

> I probably tend to be one of the more conservative people here in putting out warnings. . . . I guess if I have a slight bias, I probably would lean toward underwarning. I mean although there may be some level of minor damage with 58 mile-an-hour winds or with 3/4-inch hail, it has become evident to me that these are not life-threatening events. If I think something is just barely going to meet the threshold, I probably wouldn't warn. (Interview)

These forecasters might miss marginal "severe" storms, but hopefully they will not miss the big ones, and the public will take their warnings seriously. Through these claims, bolstered by their local cultures, they assert their authority to draw lines, transforming arbitrariness into proper professional practice. Their claim is that expertise permits an authoritative view of what weather really is.

On the Floor

I have presented an overview of life in a meteorological office, starting with the structure of the workplace and ending with the actions of workers in protecting the public. Operational meteorology, like all occupations, is constrained by the local conditions of work. Management decisions, group dynamics, and technological interfaces all contribute to what the public learns as "the weather." Organizational choices and how individuals come to adjust to and work around these constraints create work routines and public awareness.

The work of operational meteorologists involves reading the future—both an immediate future and a prospective future stretching from one to ten days. Predictive occupations differ on the time frame—the temporal window—that they are assigned as their responsibility. The structure of the weather office with its shifts, partners, routines, and warnings directs workers to organizing tasks in creating windows for action.

Because meteorological offices are continuously open, their products are not from a set of individuals, but from a team whose predictions flow into one another. Even when forecasters place their names at the bottom of the forecast, their audiences receive the work as a collective output. The public treats them as interchangeable and representative. Yet, for this invisibility they are not without their emotions—their love of weather and of "weather." Their desire is to serve and protect, to see the worst that nature has to offer, and to test themselves against this challenge. In this, severe weather is central to the way of being of a meteorologist, activating the organization and the self. Staff and structure are organized differently according to the conditions of the sky, and, thus, meteorology is not so different from other occupations punctuated with critical moments.

2

A Cult of "Science"

I am the weatherman of constant sorrow
I've seen trouble on my radar
I bid farewell to my old office
The place where I was hired and trained

For six long years I've been frustrated
No pleasures with AWIPS that I found
For in IFPS I'm bound to ramble
I have no time to work the grids

It's fare thee well my old XNOW
I never expect to see you again
For I'm bound to wait for GFE to respond
Perhaps I'll retire 'fore it responds

You can install the new AWIPS build
While I wait for years to load WWA
Then you may learn the new D2D bugs
That PRC said were fixed again

Maybe someday AWIPS will work like old blue
A face you'll never see no more
But there is no promise that is given
I'll leave before AWIPS is replaced

Brian Pierce, "I Am the Weatherman of Constant Sorrow"

: : :

Science is not a transparent window into truth. Rather, it is a *field of action,* used strategically to gain authority for assertions about the material world. Further, it is a classification scheme capable of being manipulated rhetorically; it is not a thing itself to be pointed to but an idea whose *instances* can be detailed and tied to value-laden concepts like truth and honor.[1] Scientific discourse helps some obtain resources and status and establish trust. It is not that science doesn't exist—a foolish notion—but rather that science is established collectively through the claims of organizations and small groups and is embedded in group culture. What scientists do, what they use, and what they believe—a theory/methods package—are powerfully influential. But these observable objects and acts must be combined into a conceptual framework of belief. Science is a body of acts and ideas that cohere through action and social relations.

The idea of science (like ideas of justice, equality, and much else) must be fabricated to be used culturally. I do not deny the effects of institutions or an obdurate material reality. Rather I emphasize that workers are central to scientific engagement. My goal in this chapter is to demonstrate how groups of operational meteorologists manage their status on the boundary of science as conventionally defined and as they define it.

What does it mean to be a meteorological scientist and how is this claim linked to the microculture of groups of meteorologists? Forecasters are assigned different tasks and define their occupation in various ways. Having examined three local offices of the National Weather Service, I argue that any orientation toward science and work is created by groups with their own shared pasts. Local conditions matter. By examining the office culture at the Chicago office, their impressions of other offices, and those offices' images of the Chicago office, I claim that the relationship between particular work tasks and occupational identity varies, an outcome of tradition, resources, and organizational structure.

Scientific Identities

As sociologist Chandra Mukerji emphasizes, being a scientist involves conceiving of oneself as having a scientific identity[2] and of imagining one's work as a scientific career.[3] How one thinks of oneself affects one's work, just as how one thinks of oneself flows from that work.

Much writing on scientists has addressed what some label "real scientists," workers who labor in domains in which there is consensus that their work exemplifies scientific research: academic biologists, physicists, chemists, even astronomers. Questioning this label is to question what workers recognize as unproblematic. Meteorology is one of the sciences—primarily an observational science, not an experimental one[4]—huddling under a scientific umbrella. Meteorologists may be relatively low in status[5] compared with colleagues in physics or chemistry, but academic meteorologists do not doubt that they are scientists.

Most knowledge disciplines are partitioned into academic work and applied or public domains. If the division is not quite white vs. blue collar, the status division is central. This division is tied to variable amounts of cultural capital among practitioners.[6] Whether this involves historians teaching second grade, folklorists arranging music festivals, philosophers in urban hospitals advising on medical ethics, or chemists toiling for industry, colleges do not exhaust the places of knowledge work. Science includes more than life in the academy. Transcending a focus on "research laboratories,"[7] we are only now examining the work conditions of applied practitioners.[8]

Government weather forecasters struggle with occupational boundaries, both as outside evaluators judge the matter and as the workers themselves see their position. These *operational* meteorologists both are and are not peers of their research counterparts, recognizably in the same profession, but separated by the demands of a public science, fighting to be a presence in the research-oriented American Meteorological Society, and creators of the more applied National Weather Association. Are these forecasters scientists or merely trained in science? The absence of a clear answer makes their identity a salient issue for operational meteorologists.

Sociologist Robert Merton speaks of the core of science as revealed in the acronym CUDOS,[9] standing for the values of communalism, universalism, disinterestedness, and organized skepticism. For Merton, science is created by a community of professional workers that strives to discover processes that apply everywhere, created by men and women who are unbiased and critical of unproven hypotheses. It is not that operational forecasters cannot be characterized in this way; they are, in part. But, with their focus on the prediction of the immediate future for a public audience, government meteorologists cannot engage in what they consider the *sine qua non* of science: hypothesis testing, doing of research, and publishing findings to fellow practitioners. Add to this, their recognition that they do not work in "laboratories," the spaces where scientists are found, and they recognize the problematic assignment of their identities. They challenge themselves and negotiate with others, often in jocular form, about whether what they do is science. The question is not simply "are they scientists?" but how and when might they claim that mantle.

The Demand for Science

When I challenged operational meteorologists by asking whether they considered themselves scientists, I received a range of responses. The answers were filled with exceptions, partial embraces, and hedges. Practitioners struggle with the idea of research, established as crucial to a scientific life. An extract from one interview underlines how forecasters struggle in characterizing their identity:

> Maybe [I'm] not a scientist in the sense of maybe a guy who's in a lab, and really working on equations and things of that nature, but obviously I have a scientific background. The training for a degree, the school, is all physics, chemistry, calculus, statistics. In that sense, a very scientific approach to the atmosphere. When we get into the actual job of forecasting, it sort of becomes more of a blend of science and art. . . . You're almost using scientific principles, a scientific method of looking at the atmosphere. Looking at forecasting the weather as a problem, sort of a science problem. You're taking data, you're taking scientific knowledge about how the atmosphere works. But at the same time to actually come out with a finished product, it's not just a clear-cut answer to an equation, because it's so complex and there are so many variables involved. . . . There's a certain point at the very end of the process where you've got to take a human cognitive knowledge of what's typical, what have you seen before. (Interview)

The decision of whether one is a scientist is uncertain, as the boundary markers—tied to organizational demands, work tasks, and group culture—are contradictory and hazy.

The Markers of Science. Science is often known through symbolic markers, bolstering occupational practices and communal claims. One problem is that these workers do not "look" like scientists, and their work places do not look like places in which they imagine that science gets done. The setting is wrong, and even their dress is wrong as they see it. These markers do not conform to their image of the laboratory. Forecasters emphasize that they struggle with the cultural image of the scientist, an image that doesn't fit comfortably. As one mused, "I think of a scientist with the lab coat and working in a chemistry laboratory" (Interview). Another explained, "I'm not a scientist person that works in a laboratory with beakers. Obviously I'm not that type of scientist" (Interview). A third notes, "I don't know that [members of the public] think of us as scientists. I do, but I think people think of scientists as the lab scientists and that we're not. But if you say chemists, somebody thinks of that as being a scientist. That's a problem with our profession" (Interview). They are not in a recognizable scientific space, as those spaces are imagined. Computer terminals that report what other machines have revealed do not have the symbolic potency of machines of *discovery*, such as atom smashers, electron microscopes, or even beakers.

The Research Dilemma. A second problem for operational meteorologists is the absence of research, a task that these forecasters feel is outside their duties. It is not only that meteorology is not experimental, although, as noted, that makes their workplaces problematic as science spaces, but that they do not aim for novel and generalizable knowledge. As one tells me: "I'm not a scientist in the same way that a researcher would be. I'm not involved everyday in the nuts and bolts of expanding the field of meteorology. That's not where my role is, but I do consider myself a scientist in the sense that the job that I do every day involves a good bit of science" (Interview). They are users, not dealers. They are aware that they could be doing *research*—systematic, if descriptive, analyses of weather patterns, some of which might be published in scientific journals.[10] None of the forecasters in the Chicago office use their "spare time" to compile these analyses, despite the encouragement of the National Weather Service to conduct research during their day shifts. One forecaster is adamant in claiming that, "Publishing a paper has nothing to do with our job" (Field notes). Despite the urging, significant rewards are not distributed to those who compile and analyze data. By rejecting this option, this group of forecasters places itself outside the boundaries of science as they define it.

A second element of this boundary involves the temporal organization of their assignments, and how that differs from their image of research. The forecaster's primary responsibility is to create predictions for the near future—an immediate forecast and a long-term forecast for the next ten days. These predictions contrast with research that claims to be freestanding, universal, and dispassionate. Operational forecasters evaluate models for each forecast, but for them, unlike their academic colleagues, the results do not build theory. Their findings are so local and particular as to deny generalization.

One administrator reports that his staff think of themselves as "shift workers," distinct from the temporal organization of a scientist's life. By that he means that their goal is to do what is needed during their shift, but that they lack a time frame that extends beyond those hours:

> You tend to get preoccupied with things that have nothing to do with science [e.g., learning to use new equipment and computer programs]. It has to do with the internal mechanism of the office. But one of the tenets of the job [as science officer] was to change the culture of the weather service on the floor, and have people think of themselves as scientists. That hasn't happened. They think of themselves as shift workers do, in some cases as blue-collar shift workers. Meaning that you come in, do your job, and go home. You don't have to do special projects, you don't have to get involved in things outside of this. . . . I think in this office, traditionally, since I've been here, we have very little paper publication, even write up a case or something. That is not in the culture here, and in most offices that's probably true also. (Interview)

The institutional requirements of the job—an insistent demand for a forecast at a particular moment—militates against a culture of research, and when embedded in office culture, removing the barrier is unlikely. As one put it, "you have a fiduciary duty to get this product out, so that it becomes mechanical" (Field notes). Indeed, one forecaster who expressed strong interest in meteorological theory was criticized for this esoteric passion. The temporal push of the forecast downplays academic research, or as one meteorologist told me, "it's not a college setting anymore" (Interview). Even in other offices in which some research was conducted, not all workers participated and it was not required.

The primary task is the production of assessments of the near-term future. Forecasters master new equipment for its routinized uses, rather than to use the data generated through this equipment for stable and definitive knowledge.[11] *Machines bring science with them.* The ultimate value of the machine is a matter of faith, even while a machine or program may be criticized for not working as it should. As one forecaster commented about the new generation of computers, "With modernization, there has been a huge change in the weather service. Things had really been bare bones. With modernization, we have really made great strides, bringing science into the forecast" (Field notes). He emphasizes that operational forecasters are consumers of science, not manufacturers.

Perhaps it is a function of the character of particular workers, a lack of "scientific curiosity," the demands of forecasting, the culture of operational meteorology, or the hiring criteria, but research is not essential in weather service offices.[12] The NWS attempted to emphasize the centrality of science by hiring a Science and Operations Officer at each office, hoping to provide a link to university science, but that goal has had mixed success. Indeed, the demand created some friction. In Chicago this ambivalence began as resistance toward a new form of occupational engagement, targeting a well-respected and liked science officer:

> Supposedly when [the new SOO] came here, he came here with the idea that, "Oh, I'm going to whip these guys into shape. You know, these old dinosaurs. I'm going to bring them some science, and kind of kick them in gear and get them going." And I think there was some resentment among the older forecasters at the time. It's like, "Hey, you know, we've been doing this twenty years, and this guy's going to come and tell us we're no good, and show us how to do everything right." And the whole idea of science sort of became a buzzword in that sense, and it just sort of became a joke. Not the concept. I mean obviously we fully understand that it's science. . . . The joke was more of the institution or the way that we're going to make these guys into scientists, because they don't know what they're doing. (Interview)

While my research focused on the Chicago office with its strong culture that both emphasizes and questions science, such themes are not unique to Chicago, and are evident in other offices, although in a more muted form.

Most workers think of themselves as scientists *in some way*, and then they explain the ways in which this honorific title can be justified and its limits. One metaphor by which a scientific identity is held at bay is through the claim that workers practice as much "art" as "science." They embrace the image of the scientist as *bricoleur*,[13] selecting occupational images to construct a career, none of which are *uniquely scientific*.[14] As David Laskin points out: "Weather forecasting is a science, but it's also a seat-of-your-pants kind of endeavour and an art too."[15] Meteorologists, like cooks,[16] struggle with which occupational category they belong to:

> It's art in the form of interpreting data. If you look at some of these [computer] screens out here, they have kinds of data on it. You can't tell what's there. That's fine. But that man's interpretation of the physical atmosphere, what you see out the window, is something else. How you look at a set of lines and interpret the lines or bull's-eyes or whatever, and convey that into partly cloudy, fair, snow or whatever is a leap of faith. And there's as much interpretation; you can have the same map and get three different forecasts out of it from three different people. . . . There are times when you just don't know. You can think of two or three things that can reasonably happen, but you don't have a clue as to which one. So you have to make an assessment somewhat. Right or wrong, that is the forecast. (Interview)
>
> : : :
>
> I sort of liked the fact that [meteorology] still is sort of an art. It's a mixture of science and I think we're guessing all the time. I think there's a lot of familiarity with it, pattern recognition. . . . We'd get calls from farmers out there, and they'd say, "I'm just an old farmer," but you know you've got forty years outside, and I think you have a lot better feel for the weather than I did, having a couple years experience in weather. You've got to know all the equations and be real smart, but sometimes [you] can't really put that to use in forecasting. . . . So there's meteorology, the science part of it, and the art part or the hocus-pocus part of it. (Interview)

These claims are not only a function of the interview situation but are found in the workplace, as a forecaster notes of his prediction: "Sometimes it becomes artwork. We take these numbers and massage them into words" (Field notes). These meteorologists use the art label to depict the intuitive and authentic side of their occupation, the part that they cannot specify through the rule-based knowledge of science education. They are skilled at tacit pattern recognition.[17] Forecasters skate at the edge of their scientific knowledge but still must reach a conclusion. They explain, "Now, here is the art," to justify their choices.

Being trained as meteorological scientists and located in an organization that explicitly defines itself as being tied to scientific practice, these workers sense that the identity of scientist should be paramount. Yet, as Americans who live in a cultural surround in which the concept of science conveys a set of images that doesn't directly apply, they treat the idea of science in ways to make it appropriate for their identity.

Working at Science

While the scientific self is ambivalently embraced through identity talk, the struggle for identity is also evident in action and work-related discourse. Science is enacted through workplace tasks as meteorologists examine their computer screens and discuss their forecast with partners, technicians, and administrative staff. It is not simply that they are forecasting based on the data at hand but are doing so in the *name* of science, justifying their choices by their scientific training. This occurs in a government organization in which the push to science is real, both because leaders are convinced of its benefits for forecasting and because of strategies to demonstrate accountability. If, as Warren Hagstrom has emphasized, science is a system of social control, *scientific agency* sets the agenda of work.[18]

The rhetoric of science is often utilized when forecasters differentiate their decisions from public claims or from what appear to be obvious weather patterns (what *anyone* could forecast), creating a discourse of expertise. One forecaster in preparing for the possibility of severe thunderstorms describes his caution in sending out warnings on the basis of public reports: "The biggest trick is to know the science.... You get so many bogus reports. You shouldn't jump on it if the science doesn't support it. People see things if they want to see it. The emphasis is that if the [available data] doesn't support it, don't jump out on it" (Field notes). Data and interpretations can trump eyewitness testimony. As another asserts sarcastically, "We're blinding them with science every day. As long as the masses are happy" (Field notes).

Occasionally staff make explicit reference to science. These ritual occasions are *science times*. This is underlined in an office briefing about likely severe weather; one forecaster gathered his colleagues by announcing, "It's science time." At Flowerland the MIC instituted a monthly meeting in which the SOO explained new approaches to meteorology; something similar happened in Chicago in the "Severe Weather Meeting" in the spring, the "Winter Weather Meeting" in the fall, and at sporadic lunch seminars. The National Weather Service has instituted tele-training sessions (for example, an introduction to current research on lightning strikes). Despite these training sessions, however, the only instance during my observation in which the staff of the Chicago office gathered for a "scientific" presentation was a day on which their undergraduate intern presented two brief talks about his research: one on the formation of hailstorms and the other on low visibility ceilings at local airports. While attendance was high, this was due to affection for the young man, rather than the fascination of the topics.[19]

Forecasters are sensitive about their placement in the realm of science. One forecaster complained about the office, contending that "It's hard to do original research. We've got only one journal. We have out-of-date textbooks. We don't have a library. We don't use the science. My [military] commander said, 'You're going into the weather service. Why do you want to type partly cloudy for the rest of your life'" (Field notes). Others are frustrated with what they perceive as their low status in the institutional order. Forecasters fumed about their superiors at headquarters:

> This group in power thinks we are unprofessional. They think they have to lift us up. I think they don't respect us. . . . There is real disdain for us. I think a lot of academics came into the service. I think they thought of us as the factory workers, not the engineers. . . . They came into a service organization, and they are moving it toward the theoretical part. They get excited about the science part, but the forecasting is for shit. (Field notes)[20]

: : :

> The [NWS] service is dying. [It is filled with] overgrown technicians sending out computer-generated products. Anyone can do it. I'm sick of the whole thing. People in the office doing menial work, spitting out models. I don't need it. You're stuck in this overwhelming shift work. Doctors don't put up with this. Engineers don't put up with this shift work. We spend a lot of time doing nothing. (Field notes)

Not everyone would agree with these assessments, but they are not uncommon. Headquarters, it is felt, does not appreciate that daily forecasting is a professional skill. Yet, even though these workers feel aggrieved by the lack of respect shown them as scientists, their placement within a government bureaucracy, coupled with an undergraduate education (a B.S. degree) and a heavy, routinized workload, makes any attempt to raise their status problematic. Still, the image of science becomes a central marker in discussions of how the job should be performed and what the expectations for forecasters should be.

The Problem of Operations

All workers imagine their placement in the occupation order, hoping to produce a satisfactory collective identity. Some forecasters think of themselves as physicians; they are, after all, "operational." One explained, differentiating himself from academic meteorologists, "we're the general practitioners, rather than the specialists" (Field notes). They know something about everything. Said another, "What we do and what a family doctor does are pretty similar. We observe the weather and the doctor will observe you. We'll diagnose the weather, and the doctor will diagnose you. And we'll both be doing a prognosis" (Field notes).[21] A third remarked, "The medical profession is not an exact science. They have to put things together twenty-four hours a day. We do what needs to be done to get it done. Sometimes it's not very scientific, but we're a meteorological M*A*S*H unit" (Field notes). While these informants embrace a desired occupational label, emphasizing that they are like GPs, they distance themselves from medical scientists who work in laboratories. In a society in which the evaluation of work is linked to status politics, using a metaphor of medical practice generates a warm glow.

Meteorology depends on the collection of field data, even when the field is the sky. In Jan Golinski's terms, it is a *fieldwork science*, where knowledge comes from outside scientific offices and laboratories.[22] Add to this a complex and unwieldy subject—the atmosphere, where forces are not easily measurable, and the bodies in motion are not solid—and the problem of meteorology as a formal science is evident.[23] Perhaps the odd genius might someday solve these dilemmas, but not yet. Although operational meteorologists are inordinately good at gathering and analyzing data, their emphasis that every weather event is highly contingent on local conditions lowers the possibility of replicable knowledge.

Uniqueness claims allow forecasters to escape the moral implications of predictive failures. As one forecaster explained: "We're playing the numbers game. . . . You do your best. Let God sort it out. The weather is not a perfect science. In the end, the atmosphere does what it wants" (Field notes). Such an understanding gives the skies an agency that some believe belongs in human hands. Said another, "It's very much a physical science, but we don't have a complete understanding. I've only had one forecast where I've been totally right. I was pretty pleased by that. Perfection is not easy" (Field notes). All scientific work is imperfect; there is always error, reducible but present, but the problem is more evident in meteorology than in other disciplines, particularly as the public witnesses and critiques these gaffes.

The idea of replicability in the sciences, tied to the belief in universal processes, creates a high hurdle.[24] Meteorologist Joel Curtis puts the matter succinctly as it applies to operational forecasting: "Unlike laboratory science, research, and weather forecasting case studies where the direction of the effort is defined prior to its undertaking, operational forecasting usually presents a new situation daily without controls. Besides the fluid data flow and pressure-filled environment, an essential part of the forecaster's work is problem definition . . . referred to as the 'forecast problem(s)' or the 'problem of the day.' . . . Problem identification is an integral part of the forecast process."[25] Academic meteorologists desire "doable problems,"[26] creating links to colleagues in other domains, most particularly fluid dynamics physicists. Climatic seasonal predictions are not *doable* with current models and methodologies. When an administrator was asked to appear on television to explain the long-term forecast for the following year, he tells the staff, "The odds are you stay with persistence [i.e., what the weather is like now is how it will remain], you will be right. What a joke! These long-range forecasts are meaningless. . . . I'll go down and pretend that I know what I'm talking about. The only thing I don't know is why we had this mild winter." One of the forecasters suggests, "global warming," to which the administrator mutters, "Sheesh!" (Field notes).[27] While days have routine, the excitement and the frustration for operational forecasters is that weather can be unpredictable. Forecasters embrace the claim that "every day is different."[28]

This division between academic scientists and applied or operational meteorologists led to the creation of the National Weather Association in 1975 by forecasters at the Washington office of the NWS. The split from the American Meteorological Society[29] occurred because the AMS appeared to be too oriented to academic science, showing operational forecasters "a lack of respect," and revealing a "war going on between operations and research." As one NWS employee explained, "It was a way to force the AMS to be more open to the 'operational community'" (Field notes). It is significant that he should describe his colleagues as the operational community, exposing a boundary with the academic community. One administrator commented that the division between the two groups had been so bitter that a report about the problem was titled "Crossing the Valley of Death" (Field notes). This administrator suggested—using the medical metaphor—that the decision to have the Storm Prediction Center share a building with the more research-oriented National Severe Storm Laboratory was that the SPC would be "like a teaching hospital in operational meteorology" (Field notes).[30]

The Idioculture of Science

It has been persuasively claimed that science has a "culture."[31] Whether we speak of this culture in light of ennobling values suggested by sociologist Robert Merton (communalism, universalism, disinterestedness, and organized skepticism) or other more conflictual and divisive traditions, the idea of science as a distinctive form of society, sharing norms and values, has been widely embraced.

Yet this claim erases the differences among labs, departments, and offices. A view of culture separated from interaction arenas downplays the local knowledge that structures action.[32] Sociology depends on the local. As Clifford Geertz once asserted, commenting on his colleagues, "most academic communities are not that much larger than most peasant villages and just about as ingrown."[33] Ultimately cultures are grounded in local interaction, and as a result we must focus on those spaces in which work gets done—shopfloor cultures. These group settings provide the shared basis on which work is accomplished and in which workers derive a sense of self, and such identities may differ significantly even while work tasks are similar.

Beginning with an analysis of Little League baseball teams,[34] I have argued that it is vital to conceptualize culture in interaction,[35] focusing on *idioculture,* both arising from and contributing to small group dynamics. Idioculture refers to "a system of knowledge, beliefs, behaviors, and customs shared by members of an interacting group to which members can refer that serves as the basis of further interaction. Members recognize that they share experiences, and these experiences can be referred to with the expectation that they will be understood by other members."[36] In the resonant phrase of Andrew Pickering, culture is a *mangle* in which the meaning and practices of a group are embedded.[37] Culture permits the organization of groups,[38] just as meaning grows out of the structure and interaction of groups.[39]

As members interact, the groups to which they belong develop cultural traditions. Participants share expectations of group life, and these shared patterns of interaction contribute to a recognition of belonging. Group culture incorporates traditions and practices that are tied to background knowledge, common values, group goals, and status systems[40] but also serves as a space in which new cultural items are performed that complement previous traditions. Such themes and practices, along with self-referential humorous and joking references,[41] gossip,[42] anecdotes, customs, and nicknames, characterize the group to its members and differentiate it from similar units. Participants find the essence of their groups revealed in their idioculture. These traditions constitute what Karin Knorr-Cetina writes of as "the narrative culture" of the workplace.[43] Normative expectations and values become *historicized;* they serve as a resource for achieving goals and keeping alive local experiences.[44] They constitute *moral tales* that inscribe the proper doing of work,[45] and are available for reference by members of the group. Whether we recall a remark, a prank, an object, or a narrative, the event is capable of being referred to by members of the group. Cultural traditions have what Erving Goffman describes as a *referential afterlife*,[46] suggesting that for a time a sufficient portion the group will recall the reference and invest it with meaning within the interaction order. These references establish that individuals constitute a community. Group cultures vary on the extent to which their cultures maintain traditions with lengthy referential afterlives. Strong cultures have cultural elements that recur, are referred to, and that shape the expected behaviors of group members. Weak cultures lack such elements.

In focusing on the details of small group cultures, I analyze scientific workplaces by emphasizing how group cultures shape collective identity and the work produced.[47] The examination of places and communities where science gets done emphasizes the significance of local customs, practices, and shared pasts.[48] That co-workers have mutually understood references, tied to group history, embeds their social relations in talk and performance. Scientific workplaces differ in their evidentiary cultures,[49] and also in informal cultural traditions, seemingly distant from the formal doing of science. Together these both create and are shaped by how work and identity are perceived. As a result, because of local practices, when organizations come into contact participants may feel a need for translation.[50] Scientific offices are distinctively different from each other, and these divisions matter. The form and content of workplaces shape what is produced and how. Put another way, the formal and informal arrangements of science combine to produce the conditions under which knowledge is generated, problems solved, and conflict negotiated.[51]

Because of demands on meteorologists to prepare forecasts, sometimes rapidly and under pressure, these workers develop tight bonds, even as lines of tension develop within the group. At the Storm Prediction Center, where forecasters worked in a closed room with a window providing a view from the hallway, I was pointedly told about the bond among those working "behind the glass," a fellowship so powerful that even their supervisor is "really not a part of us" (Field notes). In the case of the local offices of the National Weather Service, it is as if 122 organizational experiments are running simultaneously.

In an organization in which individuals can transfer among offices, forecasters can seek an office that fits their values—either one at which they feel content, one situated in a desirable location,[52] or one that provides them with the position or rewards that they desire.[53] The National Weather Service assumes that individuals will work in several offices during their careers, building social networks and expanding their experiences. This networking creates informal ties[54] within an organization that is built on local fiefdoms, but these linkages are spotty and based on chance affiliations. Although some speak of the NWS as "a big family," few family reunions are scheduled, and offices operate largely in ignorance of each other's practices. MICs do meet, and information is spread through their connections, a system closely tied to authority, but this does little for collegiality among forecasters.

While commonalities exist among the offices and cultures, the distinctive office culture contributes to the creation of identity.[55] As both Karin Knorr-Cetina and Joan Fujimura have emphasized, scientific practice is more locally contingent than has generally been assumed in more universalistic approaches.[56]

To examine the significance of local cultures, I focus first on the Chicago office of the National Weather Service, examining how its idioculture reveals an image of science as a central feature of occupational identity. Part of this culture involves resistance to the demands that the forecasters become more involved in research. Their collective challenge is to perceive themselves as scientists (an honorific identity), while not conducting the research that they see as defining science. The tradition of science joking is a central feature of office culture. If, as Chandra Mukerji claims,[57] affiliation with science constitutes an identity system, practitioners must announce that identity and have that announcement validated.

Chicago Weather

When the staff in Chicago answers the phone, most greet callers with "Chicago Weather," without acknowledging the larger agency to which they belong.[58] The greeting marks their autonomy. As in the case of any idioculture, the themes evident in the Chicago office of the National Weather Service are based in historical circumstances and the organizational structure produced by that history. The Chicago weather office was among the first established by the Army Signal Service in 1870. For a century the Chicago office served as the regional center for the Midwest, responsible for the forecast of several states. The Chicago office had a large oversight responsibility and considerable influence. From the 1970s for two decades, up to the creation of additional spin-up offices, the NWS operated one forecast office per state, an office with responsibility for the state forecast. The Chicago office was responsible for Illinois forecasts, as well as for regional coordination. Until the reorganization in the mid-1990s the Chicago office was still considered a regional center. Today the Chicago office has little more authority than most offices (it does forecast for Lake Michigan) with authority limited to its local forecast area. The fact that each office has essentially the same number of personnel means that headquarters has determined that forecasting for a large metropolitan area carries no special responsibility or forecasting challenges.[59] They forecast for land, not for people.

As late as the 1980s the Chicago office, then forecasting for the State of Illinois and responsible for much Aviation forecasting, had 100 employees. Today the staff has been reduced to two dozen. Some employees were eliminated through technological advances; in other cases functions were transferred to different offices. Because of its organizational centrality and higher cost-of-living, the Chicago office and a few others had a pay scale one level higher than other offices. Whereas lead forecasters elsewhere were rated as General Service (GS) 13 (the rating system on which salary is based), in Chicago they were GS-14. This status difference produced and reflected the collective sense that local forecasters exemplified meteorological competence. These forecasters felt entitled to professional autonomy, distinct from a tight bureaucratic oversight. This absence of bureaucratic control supported a culture of informality[60] and contributed to references to the office as "Country Club Chicago." The downgrading of pay scales to standard salaries still rankles some who believe that they deserved the extra pay because of their organizational centrality and accomplishments. The sense of grievance from these changes fuels current resentment, although the office's culture made complaints culturally legitimate. The Chicago office often bubbled with indignation, not far from the surface.

A Tradition of Tradition. The Chicago office maintains a "tradition of tradition"—a collective identity linked to their former location on the University of Chicago campus. In the 1950s and 1960s the university's meteorology department was the best in the nation. (The department no longer exists.) It is widely recognized, both within the office and among the leadership of the National Weather Service, that the Chicago office has a *strong culture:* a culture that is robust and extensive, affects interactional routines, and outlasts changes in office leadership and turnover of personnel. Those workers who have served in other offices find that the culture at the Chicago office is more intense. As one forecaster expressed it, "In a normal setting you would expect the culture would adapt as people bring new experiences, but this office doesn't do that. This office has a set culture in place. People that come into this office are more or less expected to adapt to that culture" (Interview). Of course, the office has a different feel depending on which pair of forecasters are on duty; some forecasters are focused, shy, or quiet, and others extroverted, sarcastic, and sociable. Even when the more gregarious forecasters are on duty, they must still focus on examining data and preparing assessments of weather patterns. The office, like many others, constitutes a place of *punctuated sociability*, in which interaction oscillates with periods of quiet and personal focus.

The salience of the office's localized tradition is widely remarked on. This was clear in speaking with a figure high in the NWS hierarchy at the time when the Chicago office was searching for a new MIC to replace their retiring leader. This NWS official explained that they were searching for someone who could respond to the office's strong culture and make changes that the NWS leadership felt were necessary. He noted that in the past the NWS had brought people into the Chicago office to "shake things up" but that these individuals had been "worn down." The office culture shaped their identities and their sense of proper practice. The eventual choice was selected from another region and had considerable management experience as well as independent ties with the Chicago media. Some months after his appointment I was told that the office had changed him more than he had altered the office.

A forecaster explained: "One thing about this office is that there is a strong sense of tradition. There's a sense of pride in it. . . . I represent my past co-workers as well as myself. It's part of the extended family of the Chicago office. Our get-togethers tend to be one big, long story-swapping event. As soon as you've been around awhile, you know the stories. I don't know if it's bonding with an office or with the people. . . . This office has deep roots" (Field notes). Many workers have nicknames, some humorous, others pungent; for example, the Beast, the Iron Duchess, Swisher, and Dr. Newshirt. Newly hired employees are taught these traditions when they arrive.

The desire for affiliation is sufficiently powerful that employees arrange communal events, including Christmas celebrations, baseball outings, a retirement party, and a marriage reception. Staff often order lunch for those who wish to contribute and have established a food "store" with candy, soda pop, popcorn, and the like.[61] In addition, coworkers gather after work, and for a time the office maintained a bowling team. This stands in sharp contrast with the Belvedere office, where no collective socializing occurs outside of the office, although employees are affable at work.

Authority and Indulgence. Perhaps as a result of the university-department model, what is most salient about the culture of the Chicago office is its resistance to authority and desire for autonomy, a point made repeatedly by both administrators and workers. Although government employees, Chicago's meteorologists believe that, like academics, they should determine their work products. Claiming professional status, they feel that their occupational culture gives them the right to make autonomous decisions. One forecaster explained the office philosophy:

> [In] some local offices . . . the MIC is really involved in all the decision-making on shift [He mentions the Belvedere office]. . . . Obviously this office is quite a contrast to offices like that. . . . Basically the forecast operations are pretty much allowed to go on with people on shift making the decisions. There's not a lot of management involved in as far as deciding what's going to be forecasted, what's not going to be forecasted. And I think that's a great thing, really. We have a lot in common that way. And as far as just management in general, this is very laid-back and very hands off. I think that's good in a sense that you're not overbearing or you don't have a micromanager that's telling you what to do in every situation. (Interview)

This attitude was dramatically evident in a discussion with a veteran forecaster. I commented that other offices are quicker to accept technological innovations from headquarters than was the Chicago office, and he startled me by referring to a comment of a former colleague, "They're good little Nazis," adding "that was [his] term for anyone who went along with the program." He added that this forecaster asked new employees if they "were good little Nazis." My informant was not suggesting that this former colleague was acting inappropriately, but rather that he reflected office culture, "There is a suspicion of the program. . . . We have a reputation for being the Central Region Bad Boys, and that's an honorable title for us." Another colleague repeated a comment from a forecaster at the Flowerland office with amazement, "We do what we're told," adding his opinion of the Flowerland staff, "Everyone's mousy." My informant added, "Here we are the kids in the back of the classroom, shooting spitballs" (Field notes). This comment was by no means atypical. One of his colleagues quoted the biologist Thomas Huxley, "Every great advance in natural knowledge has involved the absolute rejection of authority" (Field notes). On another occasion when I asked about a forecast product that I had learned that the Belvedere office distributes, a Chicago forecaster tells me that it is optional. A colleague jokes, "We just don't tell the MIC," and the forecaster adds, "What he doesn't know, doesn't hurt him." Authority rests with the staff, and not with office leadership.

Yet, these attitudes depend on the local management style. Chicago's MIC is "laid-back," providing the staff with their desired autonomy, in line with the leadership of previous MICs. He attempts to create an indulgency pattern, permitting staff autonomy, which served, in contrast, to emphasize forms of organizational control from Washington, rather than from local authority. The local office was a haven from a heartless bureaucracy.

While the MIC attempts to give staff authority, this seemingly desirable policy breaks down when staff clash. Autonomy depends on a collective agreement on propriety, the coordination of individual and group practices. Multiple perspectives on practices or on moral values can lead to discord and strife. One forecaster described this style:

> George is the type of manager who is sort of hands-off. Lets the office run sort of by itself unless large corrections need to be made, and I think there's a lot of good things about that. . . . I like the fact that I go out there and make my forecast decisions and issue my products without someone looking over my shoulder and saying, "I think you should do that," or "Why don't you do this?" Or even just being there with their nose in it. . . . At the same time, sometimes it can be so laid-back around here that there almost seems to be power vacuums. . . . George tends to be very non-confrontational. . . . He hangs back until something

> really needs to be set straight, so you have a lot of maneuvering. (Interview)

The rules were up-for-grabs in an ongoing power struggle. If each individual could act according to their preferences, conflict might be prevented, but being a collective enterprise, consistency was essential. In this, meteorological offices stand in sharp contrast to academic departments in which each faculty member sets the terms of her own work and uniformity is devalued, permitting fiefdoms in which oversight is derided as undermining academic freedom.

Even though the MIC did not insist on—and did not wish—tight oversight, there were instances in which either he or his superiors desired changes. Attempts to undercut the remnants of the MIC's institutional authority often proved successful. Early in his tenure he attempted to institute a dress code to make the staff look more professional, but the plan was blocked by several staff members and became a union issue; eventually the plan was withdrawn. On one occasion he asked a staff member to clean up a work area; after he left, a staff member jokes, "In one ear and out the other" (Field notes). The matter was not mentioned again. I was repeatedly told, as one staff member put it, "We could probably use some tightening of this office, but you would find a little resistance" (Field notes). The desire for tightening was nearly unanimous when one talked to individuals (who each had their own beliefs of what needed to be tightened), but, given the office culture, the resistance to any particular tightening would have been great. As one remarked when he learned that George would be away for several days, "When he's here, how here is he? When the cat's away, the mice will play" (Field notes). The mice desire a cat, but one that will let them eat the cheese. Although in principle staff accept limits on autonomy, they have no process by which they can collectively determine these limits. Their culture resents externally enforced limits, which emphasize that they are embedded in a bureaucracy.

The paradox of organizational indulgency is that individuals wish for more social control to constrain the "inappropriate" actions of others. They feel that these misdeeds reflect on the office and on them personally, but given the culture, nothing could be done other than to gripe. Staff held various grievances against their fellow workers: bedding down on midnight shifts, bringing children or dogs, surfing the web, putting one's feet up on the desks, or wearing jeans or t-shirts. But what was one person's concern was not necessarily another's. Too much autonomy precludes deference from others.[62] One can claim to believe in structure and yet oppose all control as illegitimate, edging the organization toward turmoil.[63]

Even after 9/11, security was noticeably more lax at Chicago than at the other offices. Tours continued, a staff member's dog often spent the evening in the office, and computer passwords were not concealed, as they were elsewhere. It is telling that, in contrast to other offices in which the radio is set to easy listening music, the music of choice in Chicago is the Rolling Stones.

With the retirement of the MIC, members of the office felt that "a major culture shock" would occur.[64] Some felt that changes would be for the best, so long as their particular pleasures were preserved. The contrast with the Belvedere office was evident; that office was structured with lines of responsibility in contrast to the more fluid, autonomous Chicago office.[65] The offices differed in their assessments of the virtue of hierarchical control as a means of organizing the routines of labor.

A Local Science Cult. The image of laboratory science is powerful for the staff at the Chicago office. The first day I observed, I was startled that the staff called one another "Doctor." I wondered if they were all Ph.D.s. In fact, most have B.S. degrees. The title has been used for over thirty years, since their days at the University of Chicago. Staff heard the professors calling each other "Doctor," and decided to adopt it for themselves, in part satirically, in part out of respect. While it is a mistake to attribute all features of office culture to their images of decades-old academic models, the image of a university department is brought up repeatedly. Under the proper circumstances, idiocultures can reverberate long after the original participants have departed.[66] Academic science provides the template of what they think they should be doing and what they know they cannot do.

This tension between mockery and reverence characterizes the local culture and directs identity.[67] Science is transformed into farce, but also provides a model through which the Chicago staff imagines their work. Even if they do not conduct research, focusing on creating and distributing forecasts and performing subsidiary office work, the image of science is central to the staff's identity. Randy explains: "The funny thing is, I think we really love the idea of science. We embrace the idea of science, but at the same time, we can sort of sit back and see the real extremes in science and sort of make fun of it.... It is not so much that we're really disdainful of it, it's just that it's fun. It's something that we share a bond. We come in here and we can laugh about something. It runs deep and it's something that we really just share in the office, and it sort of takes on a whole life of its own" (Interview).

The culture of science has been elaborated as part of the occupational identity of these meteorologists. Headlines snipped from newspapers or magazines and posted throughout the office, refer humorously to science, such as "Mad Scientist's Club" or "Science Gone Mad." The unofficial theme song of the office is Thomas Dolby's "She Blinded Me with Science." At times one of the forecasters will boom out "Science!" with an exaggerated "professorial" cadence. Several forecasters meet after work in the garage of one of the ringleaders to perform "experiments" using a microwave oven. These gatherings operate under the dual frameworks of scientific experimentation and play,[68] making evident the boundary work. One experiment—the pickle experiment—involved snipping an extension cord and placing the cut end in a pickle. The experimenters plugged in the other end and watched the pickle glow. As Randy notes impishly, "It's just an everyday ordinary pickle, and there it is glowing. What's not to like? We're simple people with inquiring minds. We want to bring about things that were never dreamed of years before" (Field notes). When Joan, the office administrative assistant, looks skeptical, Randy adds, "There might be a multitude of benefits that come to you from our reckless experimentation. . . . Someday she will look back when she is enjoying the conveniences of modern life. She'll say I scorned them at the time. I laughed at them at the time" (Field notes). They explode pumpkins with gasoline, place copper wire into sand and watch the sand coalesce, and destroy a plush toy bunny by microwaving sparklers embedded in its fur, engulfing it in flames. The scorched rabbit, nicknamed Sparky, graced the forecast computer for several days. As Joan responds, "That poor bunny gave its life for science. I'm just surprised that neighbors didn't call the police." For many weeks the office joked about a shocked pizza-delivery boy who witnessed the surprising events at this home laboratory. That several microwaves were destroyed adds to the legend, creating a mystique that characterizes their identity as "researchers." One forecaster describes the garage laboratory as "a halfway house for deranged scientists" (Field notes). Each experiment is richly narrated within the office,[69] underlining both the distance from the routines of science and the embrace of an image of scientific methodology.

Not everyone in the office participates in the joking equally—four or five forecasters are central—but all contribute amused comments. The culture is such that these themes characterize the office even for nonparticipants, who never criticize the narration. The two ringleaders established their enterprise as the (fictitious) National Center for Plasmological Research, printing NCPR business cards, creating a "training video," checking "mad scientist" webpages, and providing a collection plate for contributions. Their fantasies are given material form and provide the staff with an arena for enacting these mock identities.

On one occasion when the office administrative assistant left for vacation, they altered the lettering on the sign at the building's entrance to reflect their "new name," creating, for the moment, a cracked institutional legitimation. These two men purchased white lab coats and worked their shifts wearing these coats; one even sucked on an unlit pipe, enacting a professorial stereotype. He announced that he was considering purchasing a monocle or black-rimmed reading glasses to complete the illusion.

I occasionally felt that some colleagues viewed this play as excessive, turning the office into a house of games, but there was no attempt to control this strong culture, and most members of the staff participated, smiling or laughing at the antics. The joking was constituted as the legitimate culture of the office. Given that most of the activity occurred outside of the office, with only talk remaining, social control was never enforced. Through humor, actors can simultaneously embrace and distance themselves from attractive cultural images that they realize do not fully apply.

Odie the Imperiled Fish. The most dramatic "scientific" theme was the plot to conduct experiments on the fish that two of the female employees—the office administrative assistant Joan and the port officer Heather—kept in small aquariums on their desks. Several forecasters, notably Randy, Don, and Stan (but at least five others), teased these women that they will *do something* to the fish. The joking typically involved these male meteorologists performing scientific experimentation on Odie, Joan's fish. Joan gives as good as she gets in this asymmetric teasing relation, although the men are the instigators of the joking. We can hear echoes of gender hierarchy, despite the fact that all participants claim to enjoy the teasing. Odie's imperilment demonstrates the power of humor to allow staff to act out their imagined identities. This theme, reflecting transformed images of university science, lasted during the entirety of my observations:

> Randy and Don joke with Joan about Heather's fish. Randy jokes that her fish doesn't look well and that they should attach electrodes to the fish's tank and give it electrical shocks. Joan

> jokes that Heather will hit them if they do anything to her fish. (Field notes)
>
> : : :
>
> Randy and Stan joke with Joan, threatening to send her fish into space, burying the fish, and stapling a banner onto the fish's tail. Joan comments: "You guys are putting a lot of thought into it," to which Randy retorts, "We're scientists." Joan comments, "It's a good thing that we all get along." (Field notes)
>
> : : :
>
> As the men talk about sending the fish into space, they discuss scientific experimentation. Randy comments to Joan: "We're doing this all in the name of science. We're interested in Odie's comfort. We need to bring him back to show you the value of science. . . . We need to choose the proper rocket that gives him enough room. We need to choose the proper thrust. We're checking the parameters. . . . It can only bring publicity to [the National Weather Service]." Joan comments sarcastically, "You haven't convinced me." (Field notes)

Even as they are inverting and mocking science, scientific rhetoric permeates their discourse.

These joking sequences are an elaborate cultural creation to which all parties devote considerable energy and which they perceive as enlivening their workday, much as did Donald Roy's[70] workers when they performed "Banana Time," humorously sharing a mid-morning snack, humanizing their work environment. Through these performances it is clear that identity is a continuing issue. The elaborate plans for the destruction of the fish reveal that the workers are rhetorically oriented to a scientific identity but not to the extent that they will *really* engage in what they perceive to be academic practice. The joking keeps the theme of identity as a science worker salient. Extending the provocative metaphor of Susan Leigh Star and James Griesemer,[71] these tiny fish serve as *identity boundary objects*, revealing both the relevance and the limits of science as a defining occupational discourse. The fish reflect the tension between two images of identity and constitute the defining symbol by which meteorologists discuss—but never resolve—their shopfloor identities. Through the construction of a discourse of *experimentation* within a joking culture, forecasters emphasize that they do not perform experiments in their routine work but that experimentation is something to which they claim a personal and communal allegiance. In joking about Odie—and in other parts of the culture—they mock themselves as "brothers in science," while setting scientific practice as the sine qua non of their occupational identity. This theme is simultaneously linked to issues of status politics and gender politics, even while *occupational* politics is the ostensible reason for the joking culture.

A rich joking culture did not exist at the other offices, although each had its own traditions. What made the joking in the Chicago office distinctive and what made their culture strong was the historical and referential quality of their jokes.[72] The joking did not only respond to amusing moments but was part of an ongoing process by which the office defined itself. These fish gained meaning in this particular idioculture. That the Chicago office was particularly conscious of its own traditions and its connections with and separation from university science gave the theme power to be returned to again and again.

That this joking connected to core issues of autonomy, creativity, and rebellion allowed the theme to continue. The jokes represented the office in its own self-image. At other offices forecasters were more comfortable with the bureaucratic roles set by office administrators and by the National Weather Service. Only at Chicago, with its tradition of autonomy and skepticism of authority, did the identity of scientist as providing counter-bureaucratic resistance prove so appealing, even if it was not easily reconciled with the workers' mundane tasks. As a consequence, it was shunted into their office culture, where it became central to group life.

A Tale of Two Offices

By analyzing the strong culture of the Chicago office, I have described how a shopfloor culture can influence identity, but group cultures do more than that. They influence standards of occupational practice. Different cultures have distinct effects, even when tasks are ostensibly similar.

In examining idiocultures, it is important to recognize that cultures differ in how they permit individuals to construct distinctive local practices, even when they are connected to ostensibly comparable groups.[73] Culture shapes the expressive and instrumental contours of work. This recognition permits an exploration of how distinct values and traditions can lead to differing work products.[74] By contrasting groups in their experiences and collective meanings, the process of cultural differentiation can be made evident—what J. L. Fischer[75] labeled *micro-ethnology*. Different responses to history shape the values of small communities, and in turn those values recursively influence the behaviors of community members.[76] This finding applies to workgroups as well to small communities. Groups that appear to have a similar structure are more culturally distinctive when viewed from within.

The Chicago and Belvedere offices have dramatically different identities. The former is seen and sees themselves as suspicious of conformity, independent, and embracing autonomy; the latter is seen and sees themselves as organized, controlled, and open to institutional change.[77] Either image can be desired, although the latter fits better into bureaucratic logic, enabling it to incorporate technological innovation and submit to demands for accountability more rapidly.

The Belvedere office embraces the importance of teamwork, seeing themselves as an integral part of the larger agency. They are led by an MIC who emphasizes the importance of lines of authority and accountability. The staff, stable over the past decade, defines this as how an office should run. It is not that griping is absent but that griping is within the context of the *right* of authority to make decisions, and complaints center on whether particular decisions were correct. Many cartoons and fliers placed on walls and boards deal with management, not as in Chicago with an attempt to satirize science. The walls of the Belvedere office are plastered with Dilbert cartoons, reflecting both the commitment to management and the concern over its excesses. At one point this MIC provided each forecaster a cash award because of their collaboration when two of the staff were absent for paternity leave. The forecasters reordered their shifts at the request of the MIC. Organizational harmony is treated as crucial.

Contrasting himself to the MIC at Chicago, the Belvedere MIC emphasizes that he sees himself as a manager. He explained "I don't want to say I'm a taskmaster. On certain things I am . . . I'm probably more of a micromanager than others. I watch what we produce" (Field notes). When this man was giving me a tour of the office, he stopped in the parking lot, staring at a crack in the asphalt and commented that he would ask the paving contractor to fix the problem. He then noticed that the sign welcoming visitors to the office needed to be repainted. I felt that these were rather mundane concerns and certainly not ones that would have been commented on in Chicago. Belvedere staff, however, were willing to trade the recognition that their MIC was watching over them for the sense that he was watching out for them.

A consensus existed that the office was run well, perhaps linked to the character of the individuals who had been hired. Staff admired that the MIC was willing to fill in as a forecaster when necessary. One explained that few MICs will help out, that "it's the exception, rather than the rule" and a second noted with satisfaction that "he likes forecasting." Significantly in Chicago when the MIC filled in a shift, the staff, skeptical as they are, felt that "you wind up working both shifts." One Belvedere forecaster, correcting the MIC's efforts as his replacement, teases, "I'll have to write this up on [the MIC's] performance review.... I'm glad we have a witness to see what happens when the boss works a shift" (Field notes).

Here I contrast the Chicago office with the Flowerland office in their approaches to forecasting, or at least how those approaches are typified. The core differences between the two offices—both staffs agree—is that the Chicago office is an "old" office, once had a large forecasting region, and has forecasters who are long-time employees. Seven of Chicago's ten forecasters had arrived prior to the establishment of the Flowerland office. In contrast, the Flowerland office is a spin-up office, located in a small midwestern city, established as part of the NWS reorganization plan. Full-time operations in Flowerland began in 1995, and forecast responsibility for their County Warning Area did not begin until 1999. As a result, the organization is somewhat underdeveloped compared with older offices. The staff is still adjusting to each other and to their conditions of work. Perhaps for this reason considerable emphasis is placed on training. More training sessions occurred in Flowerland than elsewhere, and, perhaps for the same reason, there is a greater concern with personal stress: the MIC considered himself more of a personnel psychologist than a manager.

Because of the relative inexperience of personnel in spin-up offices, often located in less desirable residential locations, forecasters at the older offices frequently denigrate the competence of their colleagues at these new offices. In addition to doubting their ability to make routine forecasts, spin-up offices were seen as too quick to expect the worst and needlessly forecast severe weather.[78] Even though the Flowerland MIC pointed to better-than-average verification statistics for severe weather warnings, Chicago meteorologists considered the office too willing to scare their public. Older offices, particularly Chicago, are known as conservative and more cautious in forecasting dangerous weather conditions.[79] The Storm Prediction Center, which coordinates with local offices during severe weather, recognized distinct office philosophies;

what one office fears, making them "warning-happy," a neighboring office might see as routine bad weather, "taking them a long time to pull the trigger." Chicago exemplified this latter type, whereas for new offices everything that looks bad on the radar is assumed to be dangerous. Experience may produce complacency and confidence in one's identity, while a lack of experience may lead to a desire to act, so as to proclaim competence publicly. One of the Chicago forecasters compared the two offices:

> It's a little bit more of a likely problem in the spin-up offices with less experienced people, and that's part of it, I think. Less experienced people are maybe more gung-ho to put out warnings with less evidence. . . . For example, jumping at some model evidence three, four, five days down the road, given that models can change a lot in three or four days. Having seen these types of things happen over my whole career, I would be more conservative toward talking about heavy snow on day four in the forecast than, say, the folks who haven't worked. (Interview)

This contrasts with a competing view from the Flowerland office:

> It's a philosophy [in Chicago] that is not necessarily coming from management, but it's in the staff itself. It's ingrained in the staff. . . . Newer members of the staff in Chicago have the feeling like we do that, "Boy, I'd like to pull the trigger and do something, but the people that have been here longer and are the resident so-called experts in it are going to put pressure on me, 'well, why did you do that so fast? You should have held off longer.'" It puts a frustration factor on some of the people up there which tends to come back to us, and say, "See, they think we should pull the trigger too," like we have. But they feel they can't do it because some factions up there are going to be against it. It's a philosophy that I think that you'll see. There's a hierarchy that's built in that doesn't necessarily fit the manpower schematic of the way the office is supposedly structured. (Interview)

While forecasters in Chicago emphasize the value of continuity, forecasters in Flowerland suggest that they should call the weather as they see it. These choices are directly linked to how the group of workers perceives its responsibilities, drawing boundaries with groups with different philosophies. With the typification of experience as a marker of identity ("having seen everything"), one's meteorological choices are shaped.

The Chicago Perspective. A hidden sympathy for Flowerland culture may or may not exist, but it is clear that in Chicago idioculture often targets the forecasters in Flowerland. At the Chicago office, defenders of a less conservative approach would take some heat. In part, this may be explained by the fact that Chicago was once responsible for distributing the forecast for some of Flowerland's counties. Flowerland forecasters could then update them, which one Chicago meteorologist felt they did too readily, not showing proper organizational deference. He noted: "You'd issue the forecast at 3:30, and they'd amend it at quarter to four. That was kind of tacky to do. You're not even out the door, and they changed the forecast. You can't be that wrong in fifteen minutes. They were a bunch of hotdogs. That previous crew thought they were smarter than we were. They must have thought that we were sitting here fat, dumb, and happy" (Interview). Part of the conflict resulted from interorganizational rivalry and the threats to self that such rivalry implied.

Flowerland is currently an independent office and has the authority to forecast, but the humor, tinged with annoyance, remains. Forecasters are friendly across the borders, but within the office boundary work is evident, differentiating the two offices in their moral character and standards of good work. Part of the humor in Chicago targets what the staff derides as the "doom and gloom" and "rambunctious" Flowerland forecasts. The other part is a function of the overwrought and excessively elaborate Area Forecast Discussions that Flowerland distributes. The discussions are aimed at the local media and neighboring offices and specify the rationale for their forecasts. In one memorable case a Flowerland meteorologist mentioned weather conditions on the Kamchatka Peninsula while justifying his forecast for the Midwest.[80] Chicago forecasters sometimes measured Flowerland's written discussion by the inch. Most discussions are about 2–3 inches in length, but I was told that one of Flowerland's was 21 inches long. Humor serves as social control when a Chicago forecaster overtheorizes or overwarns and sets standards for occupational practice. One Chicago forecaster joked to a colleague, "You'll have plenty of time for your Flowerland style forecast discussion" and another remarks, "Lewis was in his

Flowerland mode" (Field notes). Discussions of the excesses of Flowerland are common, suggesting the public impact of these claims:

> Discussing Flowerland, Stan from the Chicago office noted: "They would put out heavy snow warnings. They thought the end of the world was coming. They thought there would be 6–10 inches, and we thought that there would be flurries. I guess someone told you about their 'wall of water.' A guy put in [his Area Forecast Discussion] to be aware of a 6-inch wall of water. The stupid stuff. I guess they were serious when they said it. . . . Sometimes they were so crazy." (Field notes)
>
> : : :
>
> A Chicago forecaster talks heatedly about an AFD from Flowerland that raised the possibility of an F0, F1, or F2 tornado in their warning area two or three days later—too far in the future for a reasonable forecast of severe weather, he felt. Another worker adds, "There's some irresponsibility. You're scaring people. It just convinces people that the Weather Service doesn't know what it is doing. . . . You have to weigh public need to know versus causing panic." He jokes about putting out an AFD: "There will be an F4 [severe] tornado in Grundy County on Wednesday at 2:56 p.m. The only question is whether it will verify. You can have someone go out with a truck and make circles in the field." (Field notes)
>
> : : :
>
> There is a possibility of flooding, and Greg comments, "You should look at that flooding statement Flowerland sent out. *War and Peace* wasn't that long." Don says, "I don't think they're bad meteorologists. They just have less experience. The younger forecasters are more enthusiastic. They tend to be a little excessive. They haven't had the experience that we have." (Field notes)

Even the smallest errors are mined for their humor and a means of moral differentiation. A forecast reading, "A 30 percent chance of evening showers . . . mainly in the evening," led one forecaster to jeer, "Flowerland always does a quality forecast." The typification is sufficiently strong that when another office spells "cloudy" as "clooudy," Sid remarks "worthy of Flowerland." Whenever Flowerland does not warn, the Chicago forecasters profess themselves shocked, commenting that "it's not like them not to warn" (Field notes). Yet, my field notes reveal almost as many times in which the Chicago forecasters joke about how strange it is that Flowerland hasn't warned as times that they joke about excessive warnings. This sarcasm fits the group style of many of the Chicago forecasters; they use their colleagues to enforce work norms that they do not want enforced through explicit hierarchical control.

The Flowerland Perspective. Flowerland typifies Chicago for their part, turning the tables on the older office:

> A snowstorm is tracking into the Flowerland forecast area, heading toward Chicago. Someone asks Grace, the forecaster on duty, if anyone is going to put out a winter storm watch. She responds laughing, "Nobody that I've talked to . . . that is, Chicago." The next day Flowerland had put out a storm watch, and Grace asks if Chicago had put out a winter weather watch. Bruce tells her that they hadn't, but adds, "At least they're talking about it, which is a miracle." All of the surrounding offices had issued a watch by 2:30 p.m., even the two offices to the east of Chicago. Finally Chicago issues a watch, and Jake remarks, "Chicago called our bluff. If Chicago put one out, it must be really bad." (Field notes)

: : :

> I think we're a more progressive office than Chicago. We're more in tune with our user community than Chicago is or they have been, but they have a formidable task in that regard. I think we're pointed in the direction where the weather service modernization wants us to go [more] than the Chicago office. (Interview)

By typifying Chicago as old-fashioned, Flowerland justifies its lack of meteorological experience while constituting its openness as the wave of the future.

The ultimate strategy for other offices is to play the "Plainfield card," referring to the deadly 1990 tornado for which the Chicago office did not issue a warning, using this tragedy to typify their meteorological perspective, and thereby connect their culture to the outcomes of its work practices. As one Flowerland administrator said, after recognizing their stereotype and defending their verification statistics: "Something that hangs over Chicago's head is the Plainfield tornado. . . . The offices to the west were issuing warnings. . . . There is that perception that Chicago waited too long. [The MIC] had desires to make changes in Chicago. . . . They knew how to stonewall him. There have been changes, but not that much. That's been the case with Chicago way before" (Field notes). Of course, the reason that Plainfield hangs over Chicago's head is that other offices hang it there whenever identity work is needed.

All offices have cultures that depict proper work roles and practices, and each culture is relied on to make claims about the nature of the world and their experience of occupational principles. An office's history, the collective representation of that history, and its active response to that history suggest how work should be done and how workers should define themselves. Each office justifies their attitudes toward warnings based on attitudes toward weather patterns, and then they use this knowledge to draw boundaries with those who hold different views. When linked to occupational divisions (such as between old and new offices in which the older office once had authority over the area of the newer one), these boundaries become salient in justifying a distinctive office culture as a moral domain. The combination of factors makes the differences between offices appear real and unambiguous. Because weather systems are often "passed on" from Flowerland to Chicago, the other office's decisions are salient, and, when different, can be taken as a comment on competence.

Plainfield

To examine the social consequences of disaster and how it arises from group culture,[81] I focus on a particular instance: the August 1990 Plainfield tornado. This was recognized then and now as one of the most profound failures of the National Weather Service—the mother of all blown forecasts. A meteorologist who studied the event told me, "I'm not sure we ever had a wider miss than that" (Field notes).[82] During the period of my research, the memory of the mistake still haunted the Chicago office and still, over a decade later, characterized the office.

Unlike Dorothy's Kansas, Chicago is not much troubled by severe tornadoes. As noted in chapter 1, tornadoes are rated on the Fujita scale, a measurement of the damage caused by a tornado, ranging from F0 to F5. F5 tornadoes touch down in the United States on average about once a year. At the time of my research, the Chicago area had experienced only one F2 tornado since Plainfield. Small tornadoes hit the far suburbs of the metropolitan area, but whether because of the heat island effect or the lake, the densely populated city and nearby suburbs had been spared major damage. (On April 20, 2004, an F3 tornado ravaged the small town of Utica in north central Illinois, killing eight residents. The Chicago office issued an early tornado warning, giving residents over half an hour lead time, saving lives and mitigating the memory of Plainfield[83]).

August 28, 1990 was a day that will live in the memory of meteorologists in Chicago. On that day—from 3:15 p.m. to 3:45 p.m.—a rare and climatically unusual[84] F5 tornado hit suburban Chicago's Kendall and Will Counties, killing 29, injuring more than 300 people, and causing more than $160 million in property damage.[85] An estimated 470 homes were destroyed and over a thousand homes were damaged. The storm devastated a school and an apartment building in the outlying suburb of Plainfield, from which the tornado received its name. The tornado track was approximately 16 miles long. Even several years after the event, one could see the path of the storm. The storm cell, although not the tornado, traveled through Illinois for more than four hours, originating near Rockford in north central Illinois, eventually dissipating in central Indiana. The storm was remarkably powerful and long-lived, especially for a midwestern tornado. The supercell produced five smaller tornadoes in addition to the deadly F5. At the time of the storm, the National Severe Storms Forecast Center (SPC's predecessor) had issued a severe thunderstorm watch for northern Illinois.

The Disaster Survey Report was unusually harsh in evaluating the Chicago office, despite its carefully couched bureaucratic language:

> As the storm moved into the warning area of the Weather Service Forecast Office at Chicago, the severe weather services provided were not as timely or accurate as they might have been. The first two severe thunderstorm warnings issued at Chicago were for locations that remained to the north and east of the actual path of the severe weather. The large hail, damaging winds, and tornadoes in southwestern Kane, northeastern Kendall, and western Will Counties occurred, essentially, without warnings of any type in effect. The last several minutes of the major tornado were covered by a severe thunderstorm warning that included western Will County. During the time that the supercell storm was moving through Kane, Kendall, and Will Counties, neither the staff at Marseilles [the radar facility in 1990] nor Chicago recognized severe thunderstorm signatures or indicators of the

> storm's tornadic potential exhibited by its radar echo. Few reports of severe weather, none of funnel clouds or tornadoes, were received at the Chicago office, contributing to the slow recognition of the continuing serious threat this storm posed. The lack of spotter reports and limited flow of information in northwestern Illinois prior to and during the severe thunderstorm event, coupled with the failure of radar operators and forecasters to recognize the severe nature of the long-lived supercell thunderstorm, indicates that training and preparedness activities and severe weather program oversight had not been implemented effectively at Chicago during recent years.[86]

The report contains many recommendations for improvement of procedures and a detailed account of the events of that day, stripped of the interpersonal intrigues and gossip that I learned as an observer. As a consequence of the storm one staff member was forced to retire and other careers were stunted. A lawsuit was filed (*Bergquist v. U.S.*), aimed at the National Weather Service, but because this was a governmental agency operating in its proper capacity, taxpayers were shielded from monetary damages.[87] Through a disaster, as Diane Vaughan notes of NASA's *Challenger* failure,[88] one can witness the dramatic effects of the banal, mundane qualities of organizational practices, such as the autonomy of office decision making, the structuring of lines of authority, and the control and choices of technology. What might otherwise be considered deviance becomes part of expected professional practice. But it also demonstrates the importance of an approach to interaction that attributes agency to individuals and their social relations. My goal is not to assess blame after all these years, although there is certainly blame to distribute, including to the personalities of those on duty, the interpersonal friction in the office, the skepticism about the likelihood of severe weather, the fact that important pieces of radar equipment had not been fully implemented, the cautious deference of the National Severe Storms Forecast Center to the authority of local offices, the new leadership in the Chicago office, the lack of institutional support to create a network of trained observers, the public sense that they should not call the weather service office, the number of employees, and the expectations based on local meteorological history.[89] Failure has many parents—some personal, some interactional, some organizational. Disasters are overdetermined.

What is perhaps most interesting is that the office *almost* got it right. At 2:32 p.m. the Chicago office issued a severe thunderstorm warning for *northern* Kane County. It was in *southern* Kane County that the tornado began. At 3:23 p.m., a severe thunderstorm warning was issued for southern DuPage County, just across the border from northern Will County, about twenty minutes before the tornado devastated the heart of Plainfield. The political organization of the warning system by counties served to make the warnings inapplicable. Had the forecaster merely left out "northern" in the first warning or if tiny DuPage County were slightly larger, the severe weather warnings would have been distributed for the proper area.

What was the outcome of this mistake that might not have happened and how did this forecast come to characterize the Chicago office? The events of August 28, 1990, caused considerable turmoil throughout the office and throughout the agency. All the forecasters in the office were forced to read the weather service report and then to sign off that they had read it, creating responsibility through the marker of a signature should a similar event occur.

Understandably the tornado and the weather service's role were extensively covered by the media, and, not surprisingly, the national office of the NWS was quick to investigate this failure. It was rumored that there was talk in Washington of firing the entire staff and starting anew. In an odd twist of bureaucratic politics it was the fact that the MIC had only been on duty for a few weeks that saved the office and his position. He could not be blamed for the events that transpired in the immediate aftermath of his predecessor's watch.

Accounts from those familiar with the events reveal incredulity. The forecaster on duty at the National Severe Storms Forecast Center told me, "it was not supposed to be an F5 day coming up." While they had the area under a moderate threat and issued a thunderstorm watch, no tornado watch was out. He still thinks about that day and speculates in retrospect, "We should have called up the [Chicago] office, but that was not done so much back then. We were forecasting severe thunderstorms, and we thought they've got to be on top of it. . . . In hindsight that is what we should have done." Several forecasters in the Chicago area expressed shock and disbelief at the unexpected tornado and how their colleagues missed it. Looking at radar indicators of a major tornado in late August in a community that rarely experienced severe tornadoes even in the height of the season meant that the staff didn't see it until after the fact when it was all too obvious. The future could not be seen because of the blinders of the past.

This recognition did not prevent emotional responses. Painful emotions can be heard in an account a decade later by a central figure: "[I felt] terrible, just terrible . . . I took it very personally. To me it was a blow to the office. It was a blow in prestige. . . . To me, this was a huge slap in the face. This should never have happened at the Chicago office. It was a disaster as far as I was concerned. It should never have happened. That's all there was to it. If we were doing our job, it should never have happened the way it happened, and I just thought it was humiliating. . . . And I think some of the people to this day will never forget what happened on that shift, the guys that were on shift when that happened" (Interview). One of the forecasters told me that for several months when he got home he had to have a stiff drink and suffered from insomnia; another confessed to a sense of worthlessness and guilt. He wrote poetry about the office's failure to warn their community.[90] The NWS provided no counseling for these individuals.[91] Others, not present in the office, were angered at their colleagues who let them down and scarred their identity, convinced that they would not have made the same mistake.

The idioculture of the Chicago office helps us appreciate the aftermath of the tornado. I was frequently informed that the Chicago office operates a decade later in a style recognizably similar to that of 1990, even with the change of personnel. As one forecaster commented, many of the older forecasters were "crusty." I would not describe the current staff using this label, but these forecasters were convinced of the value of autonomy, an attitude that continues.

One of the ostensible messages of Plainfield was that if a possibility of severe weather exists, it should be communicated. That "lesson of Plainfield," like that of Vietnam, is nice as an organizational summary, but, other than sensitizing forecasters, this label doesn't contribute much in that there is no doubt that tornadoes should be warned for. In my interviews I asked, "could Plainfield happen again?" and they came to radically different answers, given different meanings of what "Plainfield" was for them. "Plainfield" will never happen again (new equipment, new personnel, new forms of community), so the issue is, given the particulars of the moment, could something different but equally disastrous happen? The answer in Chicago and elsewhere is that we do not know.

Chicago and Plainfield. Despite the disaster, the Chicago office did not alter the core of their culture. Organizational cultures, whether in the NWS or at NASA, are remarkably resistant to change. While there was an attempt to close ranks in the face of a government fact-finding team, in a culture that was so grounded on autonomy, cynicism, and conflict, individual blame was readily assigned, but not collective responsibility that might cause a rethinking of their values or the structure of the office. I was frequently informed, "The wrong people were on at the wrong time" (Field notes). The complaints were focused on *who did what*, complaints that were distinguished from consideration of larger organizational issues.

Twelve years after the fact some personal echoes could still be heard in subtle ways. While many of the "crusty" forecasters have retired, I was regaled about how the culture and traditions of the Chicago office had remained stable. Although one might have expected that the aftermath of Plainfield would have primed the Chicago office to respond to the *hint* of severe weather, that did not happen. According to my informants at the Storm Prediction Center, the Chicago forecasters as a group constitute one of the most conservative offices in the nation in forecasting severe weather. Although I was told that the MIC is still "afraid that he will miss the big one" and is "a little skittish," the forecasters deride his concerns, and in these comments reveal their demand for autonomy and skepticism of over-response, demonstrating by their blasé attitudes that they have seen it all. Thus, the lesson of Plainfield is not much felt because of the power of the local values to which the office holds. Their autonomy and skepticism might prevent a disaster, or the absence of culture change and bureaucratic accountability might lead to another Plainfield. In the case of rare events, claims of future organizational failure are about as predictive as whether a tornado might hit in two months.

Yet, paradoxically a culture of Plainfield haunts Chicago. I was struck on August 28, 2001, that one of the older forecasters announced that the day was the "Eleventh Anniversary" of Plainfield. True, they didn't bake a cake, but they did remember. The forecasters at lunch described where they were when the tornado hit. These narratives were riveting. The tornado and the mention of it have a firm place in their memory, even if it hasn't altered procedures much because it was blamed on individuals and machines. Plainfield was institutionalized as an indicator of the unpredictable:

> Don comments that there might be severe thunderstorms this afternoon, and Ritchie, a college intern, comments, "Plainfield, part two." Don looks around to see who else might be present, and cautions him, "Don't say that." He then asks who will be on duty that evening. Ritchie says Randy, and adds, "He doesn't have a slow trigger finger" (Field notes)

: : :

> Lewis asks me if I have seen much severe weather during the research, and I told him that there have not been a lot of severe

> weather days since I began my research. Lewis says, "Nothing drastic. No Plainfields." Mitchell, one of the HMTs responds, "Shhhh! You don't want to provoke the Gods." Lewis agrees, "That's right." (Field notes)

Plainfield is mentioned, but with the gingerly concern that the name Voldemort produces among Harry Potter's consorts. References to Plainfield are treated with exquisite sensitivity, and, on the surface, the disaster is drained of any current relevance. Yet Plainfield is central to the deep culture of the office, signified by this rhetorical care. The office feels that problems associated with that day have not been solved and, further, that discussion of the day could rend the bonds between co-workers. Plainfield helps to define the group culture. I was told the successful forecast of the deadly Utica tornado may exorcize the specter of Plainfield, but only time will tell if a recent success can erase an historic failure, so cemented into collective memory.

The Organizational Politics of Plainfield. The restraint in speaking of Plainfield within the walls of the Chicago office is not matched by those outside. For Chicago the disaster was agentic (tied to individual choices, preserving their cultural legitimacy); for those outside it was structural (undercutting the culture and rules of practice). In times of troubles a tension may exist between blaming individuals, limiting and encapsulating blame, preserving the structure and its culture, and blaming the system, threatening structural and cultural relations. Insiders often choose a scapegoat, directing attention from a system to which all contribute.

One of the administrators of the Central Region of the National Weather Service in talking about the need to hire a new MIC explained that they wanted a replacement—twelve years later—who would insure that Plainfield doesn't happen again. Staff in the National Weather Service continue to stigmatize the Chicago office, in part because there remains resentment that Chicago was a larger office with a higher pay scale, in part because the Chicago office is perceived to be arrogant, and in part because it is perceived to be resistant to bureaucratic culture. Plainfield provides a powerful and unarguable slogan in which these concerns are combined. I was told by one long-time Chicago employee that prior to Plainfield: "I think meteorologically [the office] was trusted. I think otherwise it was not. I think there had always been a resentment to this office. People would not want to call here because they didn't want to have to deal with Chicago. . . . I think some people had great envy because we were always the higher graded office here, and I think it was the thing that they felt they were doing the same kind of work. Why did we make twenty thousand dollars more than they did? (Interview)."

Plainfield allowed this structural resentment to gain political legitimacy. One forecaster explained, "I was here and I felt [the office] was [stigmatized]. Maybe more so than before because the stigma before was a couple people who were jerks on the phone, and all the stuff was internal to the office. Nobody came in to look at it. And now all of a sudden, not only did they look at Plainfield, but they looked at all the other little things, and said, 'Oh my God! What's going on here?'. . . But it's not like those things had just happened the week before. When I started my career those things were going on and I'm sure they didn't start the day I came to work in this office (Interview)." Another forecaster explained that ham radio operators, essential for receiving reports of severe weather, still mistrust the Chicago office, feeling that "we still have the Plainfield attitude" (Interview).

I was assured in Flowerland that they would never have an event like Plainfield "because of the amount of training we do, and the amount of discussion." As noted, Flowerland staff played the Plainfield card to differentiate themselves from the Chicago office. In a staff meeting on "failure models" in Flowerland, Plainfield was mentioned explicitly but not by name, "There's the one outlier case, but to save nearby offices embarrassment, I won't mention it. Not only did everything go wrong, but nothing went right. . . . In that case, I would play up the [failure of] organization as a case that was doomed to failure from the start" (Field notes). The organizational "lessons of Plainfield" were not lost on other offices, and they were ready to mention those lessons to distinguish themselves from Chicago. Plainfield could be used simultaneously to underline interoffice differences and strengthen intraoffice community. The label becomes a cultural marker that tames the threat of organizational deviance. As part of the background culture of all operational meteorologists, Plainfield is an easily retrievable resource for the creation of local cultures.[92]

The reputation of Chicago spread widely, leading some forecasters to consider whether they should apply for positions elsewhere. In numerous interviews, forecasters told me that it still affects the reputation of the Chicago office, and I was told that regional and national administrators of the National Weather Service continue to raise Plainfield at meetings to refer to any failure of the agency. One outside forecaster put the matter dramatically, "They maybe should've just burned that office to the ground and gotten everyone out of there, and just started over from scratch, . . . because that was just ineptitude" (Field notes). Said another, "Plainfield is on our minds all the time when it comes to convective weather [tornadoes and severe thunderstorms], and so therefore we don't forget about it. Now there are other places where *absence makes the heart grow fonder. . .*" (Interview). As one local forecaster told me, "All I can say is that Plainfield lives on in other [local forecast offices] in terms of a reputation for the Chicago office" (Interview).

I interpret the response to failure as grounded in organizational culture. The strong culture of the Chicago office *might* have led to the responses to that severe weather, but it was certainly the culture of the office that led to the responses to that failure to forecast. The public recognition of that office culture contributed to the solidification of the reputation of the Chicago office, even after more than a decade. Perhaps today that culture has changed, as reflected in the office's success in forecasting the Utica tornado, but it was clear that at the time of this research, the Chicago office stood accused and felt themselves so. Ten seconds, if dramatic enough, is sufficient to stigmatize a culture.

A Cult of "Science"

Even a realm such as operational meteorology, less invested in the creation of "generalized knowledge" than in informing people if they should grab their umbrellas or head to the basement or beach, is embedded in scientific discourse. Part of this develops from the socialization of these men with their B.S. degrees. They are *bachelors of science*, seduced by but not wedded to science. They emerge from training in fluid dynamics and in collegiate work in laboratories. Science matters for their identity, but science as used, shaped, and transformed.

The identity work of these forecasters is tied to their sense of the relevance of science practice while it is simultaneously tied to the public and to the bureaucracy that employs them. These cross-cutting images create lines of strain. Workers in most institutions have their own identity challenges. Are academic researchers teachers, trainers of callow youth? Are industry researchers shills for investors? Are workers at nonprofits advocates for the politically correct demands of their employers, hoping to gain continuation of funding? That operational meteorologists have identity strains links them to others in science and elsewhere.

Although science is a form of labor, it is not something that individuals face alone, nor is it something they face en masse. In contrast, the meaning of science is connected to group relations—in the laboratory or, here, "on the floor." Work worlds have local cultures, and these cultures shape a group's orientation to the idea of science, to their successes, and, as in the example of the Plainfield tornado, to their failures. How workers at an office think about themselves contributes to how they think about their work and to the outcomes of that work. Each office is in effect a scientific experiment.

Scientific workplaces should be conceptualized as small groups, whose shared understandings result from needs for group cohesion and self-formation. These dynamics are influenced by various features, including the traditions of the group, their tasks and experiences, and the forms of social control that direct their actions. The internal character of group life is influential in the local construction of knowledge regimes. The intensity of scientific groups, coupled with the ideology of collaboration, may make idiocultures more influential in workplaces in which scientific and technical knowledge is generated than elsewhere.

The examination of scientific work has often been situated at some distance from the understanding of group life, but the reality of life in local weather service offices suggests that a tighter linkage is desirable. Idiocultures affect both identity and the conception of good work. While some elements transcend group boundaries, the ability of a group to provide a cognitive and emotional structure[93] gives groups a normative power.

An office can develop a strong culture that situates workers in particular locations within the occupation to which they ostensibly belong. The comparison of the Chicago and Flowerland offices emphasizes that an office culture affects work outcomes, including matters of public safety. Residents within the Chicago and Flowerland forecast areas will learn different things about their weather and will be given different advice and warnings as a function of contrasting office cultures. With 122 offices, where one lives affects the kind of weather one receives, not meteorologically, but culturally.

Groups form not only within the laboratory, but wherever workers congregate. Occupations consist of webs of social relations that tie individuals to tightly knit networks. Groups that share space (colleagues in offices or laboratories) and those that don't (invisible colleges of distant colleagues working on similar problems) develop practices that remind participants that they belong together and are separate from others.

3 Futurework

"We live on data, and it's highly perishable. Yesterday's forecast is not worth anything."

Field notes

: : :

Weather forecasting trades on the future. Yesterday's forecast, like yesterday's weather, is a curiosity, but tomorrow's forecast matters. Despite public complaints and jokes, the National Weather Service is not only one of the most respected agencies of the federal government but also one of the most frequently used. Reports suggest that a large majority of Americans access weather news on a daily basis, relying on it to plan their local futures.[1]

Meteorological forecasts have consequences. Perhaps as a result, the chair in which the public forecaster sits is called the *hot seat*. As one meteorologist put it, exaggerating slightly: "We make it up. You look at TV; they are regurgitating what we put out. [You have no responsibility] unless you're sitting in the chair and have to decide: Do I do it or do I not do it? If you blow the weather, you affect a lot of people and a lot of money" (Field notes). While this claim boosts their legitimacy, particularly in contrast with the media,[2] it emphasizes that the forecasters at the National Weather Service see themselves as a bulwark against natural disaster. The fact that forecasters do not describe what is but rather what *will be* makes them of especial interest to those who examine the contours of advice.

Social systems are Janus-faced. They peer intently forward at the moment they gaze back. Yet, social research examines the latter at the expense of the former. Futures research is a domain unto itself, distant from sociological analysis.[3] Given the availability of secure data, it is understandable that there should be historical sociology, but not futurist sociology. More surprisingly, researchers have barely examined the process by which we prepare for—and make claims about—the future. Most social systems, however, establish ritual specialists whose responsibility is to explain what the morrow will bring. Predicting and preparing for that which has not arrived is of considerable significance in permitting social order, giving confidence in coping with change.

Whether we consider physicians, financial planners, actuaries, pollsters, fortune-tellers, or meteorologists, social systems have invested a set of occupational workers with the task of prediction, smoothing the path to the future, allowing us to structure the present to prepare for what we imagine is to come, and in the case of futures with feedback loops, potentially to alter that future.

A core threat to any social system is uncertainty.[4] Misfortune is easier to cope with than ambiguity,[5] and so within social institutions we discover an "effort after meaning."[6] We seek predictability in planning our lives. The vigorous dissemination of claims of what will transpire represents an attempt to create collective expectations—a shared future.[7] We strive to narrow the range of possibilities—harmful and beneficial—to a single interpretation, creating the possibility of collective action. Even if we do not consciously accept this exclusive future as the only one, we typically act upon it.

I hope to understand how forecasting is a social process[8] by addressing the conditions under which claims of future events are advanced.[9] Perhaps social scientists should make future predictions themselves,[10] but regardless, we surely have a mandate to examine how predictive claims are created and become persuasive.

Predictions can address futures at any distance, near or far. While there are similarities no matter the time frame, differences exist as well. I address short-term futures, examining the work routines of operational meteorologists, whose occupational task involves the production and then distribution of accounts of weather systems for the following ten days. Considerable distance separates these workers and those who present prophecies for the remainder of the twenty-first century.

Yet, both groups make claims of events that have not transpired, and both require a system through which such predictions are generated and announced.

How is the future organized as an occupational routine? How can prediction be linked to life on the shopfloor? It is not necessary that these workers be correct—they are often wrong—but their occupational legitimacy depends on the belief among their audiences that what they suggest could likely happen. Audiences must have sufficient confidence in the claims that they will act upon them.

Four elements are necessary for a public prediction: gathered data, disciplinary theory, historicized experience, and institutional legitimation. These categories are not entirely distinct,[11] but they serve as organizing concepts for transforming past to future. Each has its limits as a window into the future. First, the predictor must acquire empirical data, using a variety of technological devices, constituting a base from which extrapolation is possible. The collection of data results from institutional politics, resource allocation, and technological choices. These data are not transparent and must be translated and massaged to become useful for the forecaster.

Second, the predictor requires a theory, grounded in a knowledge discipline, that permits routine strategies of extrapolation from the available data. Models of knowledge are essential for these claims. Theories serve as a routine basis from which current data are extrapolated. They bring scientific legitimacy to the task of forecasting, suggesting a tested and proven basis for predictions. Of course, the predictor can ignore or alter these assumptions, as they pose status challenges for those with the responsibility to make forecasts. Theories are generated by specialists in locations outside of those places in which local forecasts are routinely made, and as a result they must be invited into the forecaster's world.

Third, forecasters as a community of professionals[12] base their forecasts on the primacy of authentic experience. Even if daily weather conditions are always special cases, and thus idiosyncratic, forecasters historicize data and theory through the claimed similarity between past and present. Workers have encountered weather patterns either directly or from their occupational community, establishing trust.[13] As a result, they treat these experiences as creating knowledge that transcends both the seeming precision of current data and the elegance of disciplinary theory.

Finally, a prediction must be institutionally legitimated.[14] This legitimation does not affect the ability to *make* the prediction, but speaks to whether the prediction will be taken as valid.[15] Audiences accept claims that they cannot check, even when such forecasts may be bleak or troublesome. I refer to three forms of legitimation: one situated within the domain of specialized knowledge (*occupational legitimation*), the second tied to the institutional structure (*organizational legitimation*), and the third linked to impression management (*presentational legitimation*). If forecasters are not always confident of their pronouncements, they have to perform in such a way that others believe they are.

A body of literature specifies the conditions under which predictions occur.[16] Although this literature has lacked an ethnographic component, it emphasizes the centrality of path dependency, recognizing that the data from which forecasts are made result from past choices and these choices are treated as the way that things are always done. A recognition that these data result from choices permits us to understand that just as the future is contingent,[17] so too is the forecast of that future.

Occupations differ in their temporal orientation. Some occupations claim the past as their institutional domain, such as police detectives determining how a crime occurred, pathologists conducting autopsies, or archeologists searching for clues to the development of a society. Other occupations focus on the here-and-now. Cobblers, actors, and car salesmen are examples. And a few are given the assignment of looking forward, such as physicians, financial planners, fortune-tellers, pollsters, and, here, meteorologists.[18] They engage in *futurework*.

These boundaries are not hard and fast, but a matter of emphasis. A police detective might be charged with determining the location at which an arrest can be made or may be asked whether a criminal is likely to strike again; the car salesman, about the repair history of a vehicle or the likelihood of a car needing repairs in the future; the internist may be asked about the meanings of past symptoms, current medical interventions, as well as the patient's prognosis.[19]

We visit the doctor to learn the course of our ailments, the fortune-teller to explain the course of our affections, the planner to better our finances, and the meteorologist to plan our activities. Although not central to their work responsibilities, there are many professional occupations (attorney, priest, therapist) in which individuals demand to know what is to become of them. In each case, practitioners must be confident in their vision of the future and then figure out how they can best communicate that future, based on their persuasive skills and their interpretation of our needs.[20]

Futures have variable trajectories and different lengths.[21] As a result, we can speak of *temporal windows*. The dimensions of the window affect the strategies by which data and models are gathered and interpreted. For the operational meteorologist, the future stretches for ten days; after that the task of prediction is handed across occupational boundaries to the climatologist.[22] With the exception of severe weather in which immediate forecasts are common, the future begins two-and-a-half hours after the forecast. The 3:30 p.m. forecast formally takes effect at 6:00 p.m. In contrast, for internists and fortune-tellers the future has a longer stretch. In dealing with chronic conditions, the prognosis may take years to reveal itself, although acute illnesses have a smaller window. Fortune-tellers are rarely asked about what will transpire on the drive home but are expected to forecast the trajectory of a life with regard to affections, health, and financial windfalls. Financial planners have an even wider window—through retirement and beyond.

The Forecaster's Enchantment

The dark heart of prediction is defining, controlling, and presenting uncertainty as confident knowledge. *To forecast* is to strip uncertainty, responding to the demand for surety, eschewing ambiguity.[23] Forecasters trade in confidence as much as facts and, as Trevor Pinch[24] argues, debates in science are often fought over the degree of certainty to assign claims. Certainty belongs to the rhetorical armor of professionals; uncertainty indicates the presence of risk.[25] Forecasters must persuade themselves and then persuade others that what appears indeterminate is in fact knowable. But for those who attempt to make claims about what has yet to occur—futureworkers of all stripes—such contentions are problematic in that one can only point to the *process* of knowing as providing confidence.[26] Too much transparency may weaken authority as the limits of a field of knowledge become readily apparent.

The demand to predict under conditions of uncertainty creates strategies of claimsmaking, organizational legitimation, and semantic hedging.[27] In judging the future the glass is dark and murky, but must appear sheer and lucid, as much a political and psychological process as a technical one. This provides for occupational challenges and techniques for generating confidence in the face of doubt. As a meteorologist sighed: "That's the biggest thing about our job . . . the uncertainty. If it was completely obvious, anyone could do the job. You just sit there and doubt yourself. Everyone is doubting you because it is just not happening. Even you start to doubt yourself" (Field notes). One administrator noted of the confidence of young staffers, "You can tell it's early in their career. They are still confident in their predictions" (Field notes); their self-confidence has not been buffeted by the realization of routine error. Socialization to the limits of knowing is critical to professional training.[28]

Yet, uncertainty must be distinguished from ignorance. Uncertainty is only possible when coupled with knowledge.[29] Forecasters must recognize their boundaries of knowledge and strive to extend those boundaries by means of the interpretation of data, the application of theory, and claims of experience. They then convince audiences that these predictions are to be treated seriously and acted upon.

Although certainty may be a goal, at times uncertainty provides rhetorical distance from error and provides an escape for one's audiences, instead of embracing an unhappy certainty. Too much certainty leads audiences to recognize error, perhaps making trust more difficult later. It preserves hope in the face of tragedy. In medicine a professional rhetoric of the uncertainty of a diagnosis is evident when doctors are constrained to provide distressing or fatal news.[30] Because the future hasn't arrived, such wiggly claims are not considered a breach of ethics, but humane honesty.[31] Professionals persuade themselves that such ambiguity is understood and appreciated by their audiences. As one meteorologist noted: "People may demand hard and fast answers from forecasts, but they know that the world is often unpredictable. They just need a little reminding,"[32] a belief that excuses error.

Gathered Data

Operational meteorology is awash in measurements, and these measurements are the basis for predictive claims. As weather future is typically a slowly changed version of weather present, the data gathered are typically close enough—close enough for government work. Weather forecasting couldn't be practical until a network was developed to collect atmospheric data in a systematic fashion. That was the value of the Army Signal Corp as the basis for forecasting, and the value today of mechanical readings that can immediately be distributed to forecasters. It is through data collection that weather forecasters create the past and the present, and from this creation the future is built. The decisions of which machines to build and how to use them mean that when we talk about futurework, we are also talking about choices about how to know and understand present and past. Indicators of the meteorological present and past are to be gathered because of the need to create a future, just as the future rests on the shoulders of present and past. If meteorologists distinguish between hindcasts and forecasts, social scientists see them as mutually dependent.

Collecting data is insufficient, however; to be useful for forecasting data must be organized. And so, data collection amounts to a joint activity of groups and of machines. The infrastructure of science permits the world to be described numerically and systematically, a result of the politics of disciplines.[33] Technology must be built and understood to permit desired tasks to be achieved,[34] providing for an "ecology of knowledge."[35] To produce the future, we must produce records of the present.

Bruno Latour and Michel Callon[36] have famously argued that machines have agency—power—over the creation of scientific data in theory-method packages. Machines and their men compete with men and their machines. Are machines calling the shots or are their handlers in charge? Where does authority lie? Lacking devices to produce data, humans would be distinctly limited in what they could know. For example, "measuring the temperature" is essential in determining future temperatures, but without trust in thermometers, we could not be confident in our judgments of the temperature (the ambient "feel" of the air). The machine and forecaster become what Ivan Tchalokov terms a "heterogeneous couple."[37] The point is deeper than questioning the validity of the human senses to measure temperatures correctly. Our measuring units (for instance, dividing the range from freezing to boiling by precisely 100 to get the Celsius scale) results from aesthetic preferences; metrology—the science of measurement—is a very social science. Degrees, invisible and imprecise to our tactile senses, demand equipment to become data, creating a *visible* and *usable* marker of heat.

Machines, notably radar systems, unveil the hidden, or as Dennis McCarthy, then the MIC in Norman, Oklahoma, put it, "We see things we always knew were there, but couldn't see."[38] As Barry Barnes, David Bloor, and John Henry assert, "As the word is used by scientists, the category of 'observable' things is far broader than the list of things that can be seen, heard, tasted, smelled or touched. To 'observe'—in a laboratory context—does not generally mean simply bringing the sense organs into play. . . . The language of 'observation' is thus contextual."[39] Under scientific regimes of knowledge, we do not need to confront a natural object *as it is, where it is,* and *when it happens.*[40] Capturing data, we use it for *what it will be*. Equipment tames and transforms the world, but by this we are tamed and transformed. Our guesses are educated by machines. The use of technology is ritualized scientific practice, shaping the meanings of our world.[41] As technology changes, workers continually adjust the conditions of work. To do *good work* scientific workers routinely face what they conceive of as radical restructurings of proper practice.

It is technology that the public imagines and admires when they think about scientists, certainly in the beliefs of meteorologists. When I asked what the public would be most impressed by if they visited a weather service office, forecasters repeatedly named the upper air balloons, the radar, and the computers. As one said, "all the equipment. Everybody likes bells and whistles" (Interview). The bells and whistles, as this forecaster phrases it, persuade the public that science is happening—that predictions are more than personal preferences. Technology doesn't just produce data, it validates it and makes it worth paying for. Equipment justifies the scientist.[42]

Given the centrality of data to forecasting, I was surprised how often machines didn't work. Of course, most pieces of equipment operate as expected, but at any given moment something was always not functioning properly—the computers, the radar, the radio, the observational devices, the power generator, and many other items, including such mundane technological systems as telephones, light fixtures, and toilets. An administrator admitted that at any time about one-third of the observational stations were not gathering data properly. Each office was assigned two other offices to provide back-up services and forecasts in case of unplanned breakdowns or planned upgrades of technological systems. In addition to major breakdowns, minor glitches were frequently encountered. On these occasions, forecasters personified the detectors,[43] suggesting that the machines "burped" or "farted." An important skill is to be able to *persuade* the machine to work.[44] As Chandra Mukerji notes, scientists were often at the mercy of their machine-tenders, who are otherwise marginal.[45] The Chicago office had a particular problem when the employee in charge of the equipment was critically ill for several months. On a day in which the power supply broke down and severe weather threatened, this man was described by a colleague as "the font of all knowledge. If something were to go wrong in the backroom, no one would know what needs to be fixed." The office had to rely on backup offices for several hours, trusting their predictions to others.

The problem of control was as acute for software. Each system upgrade—of which there were many—introduced bugs, glitches, and incapacities. As one forecaster put it: "For every one that they fix, they screw up something else. Getting it fixed never occurs, because when they fix something, they break something else" (Field notes). I was always treated to a rich and theatrical performance when the staff installed new software, explaining in exquisite detail how it didn't function properly, complicating seeing the future. The staff treasured such stories, as they provided a reminder of the limits of both machines and bureaucracies and validated the authenticity of local knowledge. In time, these problems were solved, but through their culture of complaint,[46] workers assumed the malevolent—or at least mischievous—agency of technology and bureaucracy. Machines are simply another hurdle that must be overcome.

Technology as Wack. Perhaps because technology is the window to the future, it is a central point of the culture of complaint. In part because of the frustration engendered by routine equipment problems and in part because the primacy of the machine poses identity threats, resentment is common in technology talk. These complaints are based in a realization that *things* have autonomy in establishing forecasts[47] as workers cannot *start* analysis without data from these machines. As is often the case when emotional display is illegitimate, irritation finds expression in caustic humor:

> Forecasters denigrate the software of their AWIPS [Advanced Weather Interactive Processing System], remarking "This is crapware at its finest. They have actually made it worse. . . . Who actually writes this garbage? Crapware abounds." Marty responds sarcastically, "Now, now, this is the finest system that money can buy." After learning that the software won an award, they start referring to it as "Award Winning Crapware" or AWCW. In addition to describing the system as "crapware," they also routinely refer to "G.D. [god damn] AWIPS." One forecaster jokes, "With AWIPS nothing is ready before its time. Like fine wine, except it's not fine wine. It's more like moldy cheese." An administrator responds: "They're not deliberately trying to mess us up. People are howling to them to get it out, so they rush and people complain about the errors." (Field notes)

By desacralizing the machine, forecasters can justify their predictive errors, and then figure out how to work within the constraints the machines set for them.[48]

Observational technologies are not transparent windows on the world. Data are ambiguous. Again meteorologists return to the metaphor of medicine:

> It is like a doctor reading an x-ray. When we look at the radar we're not exactly seeing. We kind of have to probe the gray area. That's exactly what the radar does. It's not giving us an exact picture. It's giving the radar's interpretation of what the radar

> can see. The radar can't see everything. A different angle, it will give you a different view. Like these CAT scans, . . . one view might give one and another view might get a totally different picture than the previous view did. That's one thing you're limited to on the radar. . . . You're limited to one view. (Interview)

Forecasters must expand the data they receive, making two dimensions three, and then must extrapolate from the images, predicting ugly weather on beautiful days. What I saw as indistinguishable markings must be differentiated. For instance, Dean indicates a satellite image of a storm and points it out to my unseeing eyes, "Look at that. Beautiful. You see the shadow from the overhanging top [of the storm cloud]." He points to a radar image of a hail spike, the largest hailstones in a supercell, and a ball at the end of a supercell that indicates debris from the storm's destruction. Dean comments, "These are interesting clues if you know what you are looking for" (Field notes). These claims made sense to him and to his colleagues, even if he lacks direct access to the world as pictured. To a trained eye an image reveals not only the primary information but secondary or tertiary features that help to explain what is occurring, sometimes including dynamic processes that reveal the future. Even in ideal circumstances, the forecaster cannot see the tornado—nor does the machine; they find indicators of air circulation and the traces of cyclonic winds. It is, in effect, detective work.

The Doppler radar, for instance, is constrained by the linear nature of the beam versus the curvature of the earth and as a consequence may not detect smaller tornadoes that develop below the beam from the underside of clouds. These tornadoes are in effect invisible. As one public officer emphasized, "the radar can only *tell us what it sees*" (Field notes).[49] A lower beam, I was told, could not be "sold" because of public fears of radiation. Cancer trumps fears of twisters. New thermometers are less accurate than older ones, because of concerns about mercury poisoning, enough potentially to alter our understanding of the direction of climate change. As one meteorologist put it, "We're less accurate. Perhaps because of safety reasons, because a lot of thermometers we used were mercury which are very accurate, we are taking as many as we can out of the system, because they're dangerous, or at least it's been determined that the mercury is dangerous. They'll have to use electronic" (Interview). Automatic observation systems are not recalibrated often, and so error can creep in. Even radiosondes—upper air balloons—have their bias. As Jeff Rosenfeld, the editor of the *Bulletin of the American Meteorological Society*, phrased it in an editorial provocatively titled,

"Data are Not Data": "Rose is a rose is a rose, according to Gertrude Stein. Unfortunately, you can't say the same about radiosondes. These familiar balloon-borne instruments haven't changed much over time: that's one reason they are still the backbone of the upper-air observing system. But... the slight variations from radiosonde to radiosonde have had huge implications for climate record keeping. Little shifts in bias—good and bad—matter a great deal."[50] I cannot ascertain that Rosenfeld's concerns are justified, but the data are used as collected, without routine quality control. The extent of inexactitude is itself imprecise.

On what basis can detectors be trusted and hence be usable.[51] Following Harry Collins' discussion of the "experimenter's regress,"[52] Barnes, Bloor, and Henry assert, "discussions about when a piece of apparatus is 'working properly' can materially affect what is counted as the outcome of the experiment."[53] This evaluation of good data depends upon expectations developed in the course of processing information:

> I am sitting with Phil as he observes the data sent back by Flowerland's upper-air balloon (it sends data for 2–3 hours). Phil massages the data, eliminating what he terms "wrong data." Every so often, he says of temperature data or humidity, that the data is wrong and that "the balloon didn't do that." At one point Phil remarks about data that doesn't make for a smooth temperature curve: "That's not good. We'll take that out." He prefers extrapolation to "bad" (unaesthetic) data, even though the bad data, like the good data, were received from the detector. Phil tells a visiting student, "We do massage flights from what are considered machine errors. You are supposed to smooth out the irregularities." (Field notes)

: : :

> Diana is working on the Surface Analysis desk at the Hydrometeorological Prediction Center. She examines the data that they receive [based on observational reports], checking that they "make sense," and then distributing maps with the data. These data are the starting point for forecasting. She comments of the senior duty forecasters, "Every so often they will come over and say, 'this data is not good.'" Typically those data are excluded, but judgment depends on what seems right. (Field notes)

Accepting data is tied to assumptions of their proper range. Data outside of the range are eliminated unless their peculiarities can be justified.

The output of machines is evaluated by the limits that human actors set for them.

Black Box Orientalism. If machines are Others—present, potent, and mysterious—human agents must come to terms with their peculiarities. To gain control, forecasters personalize the machines. They can determine which types of data they prefer to display on their monitors. More saliently for an observer, they select preferred color schemes for isobars, clouds, precipitation, and the like. For some, green images refer to "anything that involves moisture," for others clearing skies. Colors do not reflect the atmospheric conditions but are chosen from cultural preferences.

Despite a rhetorical mistrust of mechanical authority (metalicism!), the cognitive default is that what is reflected on the machine is what is real in the skies. Not only do changes in technology alter how an office is organized, they alter how inhabitants think and what they consider knowable. The movement from "handwork" (plotting data by hand) prevents the forecaster from directly questioning the data.[54] This is a process of deskilling, even while the forecaster becomes increasingly skilled at the technological interface.[55]

Technology channels the possibility of knowledge creation; it smudges and hides as it reveals and opens. This is particularly evident in meteorology because the need for collaborative data collection across extensive areas depends on systems for manipulating large amounts of information. One administrator emphasized that "Every major advance in meteorology has been as a result of a technological breakthrough elsewhere" (Interview), including the telegraph,[56] cartography,[57] radio, satellites, radar, and computers. But each forecasting breakthrough produces challenges in its wake.

This is not to suggest that the latent costs outweigh the manifest benefits, nor does it deny that fixes can be found for technological glitches. Yet, technology does not make the material world transparent. The world as presented must be transformed. Machines, in effect, change our assessments of the weather.[58] They demand that we *use* the technology, justifying government funding. The cost of machines makes big science an appendage of the state.[59] This helps explain the increase in warnings after the installation of the WSR-88D radar in the early 1990s—not a result of the images themselves, but of the need to demonstrate the value of the images. It is not (only) that the data helped them see the future better, but that they acceded to the assumption that they could see that future better.

Given that change is endemic in equipment, in the software, in the algorithms, and in the desired protocols, techniques of machine tending must be relearned. Retraining becomes a central trope for meteorological life. One forecaster, active for twenty years, explained that he had to learn to forecast three times as technology changed from teletypes to composing written forecasts on computers and to forecasting using databases and numerical grids. He is still relatively young, so additional changes are likely.

My point is not to suggest that machines only cause trouble. Such is not the case. The various types of data that they provide are critical for the forecasts on which we have come to rely. And as the data becomes more precise, the forecasts are better and are better for longer periods. The data matter. Yet, this is not a simple process of naively accepting what machines provide but of understanding as competent professionals what these data mean and how they can be used.

Disciplinary Theory

Both routine and severe weather forecasts depend on data, but data do not speak until rules of interpretation give numbers and images meaning. Each science, whether basic or applied, develops concepts that serve as translation devices. Theory construction is essential to make sense of the natural world, but the theories have been proposed precisely because scientists believe that they result from a close observation and examination of that past. If they appear to be predictive, it is only because scientists claim that they are grounded in a closely examined history. They emerge from tests of previously available data.

Within operational meteorology the crucial interface with theory takes the form of models, developed from large batches of meteorological observations, whose regularities are now revealed through large sets of equations that simulate the dynamics of meteorological systems. In this meteorology is similar to other disciplines, scientific and behavioral, that attempt to predict complex, dynamic systems, such as economics or ecology. Much of what can be said of meteorological models is true of models elsewhere.

For meteorologists to ignore these theories, however distant they may be from life on the floor, is to question academic science. As in the case of data, we start with the assumption of the value of these theories and explore how they are used by practitioners in their predictions. The fact that we can be reasonably confident of forecasts for two to three days in the future is a function of the development of sophisticated models that permit the extrapolation of current data.

Embracing the Supermodel. No forecast is produced anew. Not only do forecasters rely on data and on inherited forecasts by colleagues, but several times each day they are provided *guidance* through a set of models of the atmosphere. The model is a numerical and graphical representation of the solution of a set of mathematical equations, based on theories of meteorological dynamics. These numerical weather prediction models were first developed after World War II with the growth of electronic computing, particularly at Princeton's Institute for Advanced Study Meteorology Project. This project had considerable investment and direction from the United States military.[60]

To the extent that weather systems are lawful and the theory behind them properly specified, the models should accurately forecast changes in these systems, recognizing that some dynamic systems are easier to model than others.[61] Of course, models depend on the data that they are given—how they are *initialized*. If data are erroneous, even slightly, the predictions can be askew:[62] garbage in, garbage out. Local forecasters cannot directly evaluate the accuracy of the empirical observations, except by contrasting it to what had been expected from models or experience. Likewise academic theory cannot be challenged from inside the forecast office. Errors of hypothesis have been erased.[63] Hypotheses have been modified to fit disciplinary preferences. When predictions are wrong data gatherers blame the theory and modelers blame the data. Operational forecasters blame both.[64]

We embrace graphical images of models, part of the human desire to transform information into pictures,[65] while simultaneously opening nature to theory.[66] The relationship between models and reality results from how models are created and reality is known.[67] As Barnes, Bloor, and Henry note, citing Mary Hesse[68]: "Models and theories in science are not sets of postulates from which deductions may be made to test against observations, but rather are resources for the metaphorical redescription of what is being observed or experimented upon."[69]

Guidance is a charming, deferential term that seemingly preserves the forecaster's autonomy, creating a man-machine mix.[70] These models are not to be taken as *forecasts*, although there are predictions embedded in each. The model indicates what the theory suggests is likely to happen at various places and points in the future. The distinction between this and a forecast is rhetorical, providing the human cover to enable her to be considered the forecaster. To this end, the forecaster is given a set of models, rather than a single one: it is his responsibility to select from among six to twelve models.[71] Models have different time frames and cover different geographical areas. Basically models come in two flavors: short-term models and models that forecast longer periods (up to fifteen days). The long-term models examine larger weather systems and have a broader scope, often the northern hemisphere. The short-term models, forecasting for two to three days, rely on data from North America. Because weather systems typically move from west to east in the northern temperate latitudes, midwestern weather depends on weather in Canada, the Pacific Northwest, and the northern Pacific Ocean. Although weather services of the advanced nations create their own models—the Canadian model, the UKMet, the European model, most NWS forecasters used one of two American models,[72] the ETA (and Meso-ETA) model, which focuses on short-term weather, predicting up to 60 hours ahead (now 84 hours), and the Aviation model (now called the GFS model), which predicts for 120 hours. Models can depict the broad sweep of weather systems but cannot yet forecast the microdynamics that characterize local storm cells. In meteorological terms, models are highly effective in forecasting on the larger, synoptic level, but are less effective on the more local mesoscale level. Likewise at the distance of a week or ten days the forecasts, while precise, are often precisely wrong. As one forecaster put it, "It's only marginally better than flipping a coin" (Field notes).

Complicating the use of models by operational forecasters is the fact that meteorological theory is based on seasonal dynamics; some models change their equations several times a year. Further, models are continually being altered while keeping their name. These changes hopefully improve forecasts but require forecasters to relearn their peculiarities. Models were described as being like George Washington's axe, "with five handles and sixteen heads. It's still George Washington's axe, but there's absolutely no resemblance to the original incarnation" (Interview). In this way, a model continually changes, remaining the same.

Although I focus on how models are used in local offices, I spent a week at the Environmental Modeling Center in Camp Springs, Maryland. These modelers take their work seriously and invest themselves in their models; they speak of their models as their children. One modeler explains:

> A lot of processes around here are akin to giving birth. First of all, birth is a very intense experience. Working on a computer model to produce or simulate or replicate nature is equally intensive. Many people when they've finished a project here go

> into a postpartum depression because it has been so intellectually enervating to do this, and they don't want to do the project ever again. It burns you out. People think of their models as their children. "I've done this, and this is the best thing around. It's my child. Nothing is better than my child." We have a number of examples on this floor right now of people who don't want to let anybody touch their model.... When model development gets to a certain relatively mature phase, and it looks like it's actually going to be useful, the branch takes it over. So it becomes the branch's model, not a person's model, but the person who built it has to adjust to that. (Interview)

A tension exists between those who create models, and those who focus on the analysis of observational data: Theory people and data people. As is often the case, spatial boundaries are involved. In the National Center for Environmental Prediction, theorists have offices on the second floor, while those who analyze data are situated two floors above. As a result, the division is described as between the second and fourth floors.

While the two groups collaborate, the relationship can be sticky. After all, if the theorists are successful, there may no need for human forecasters. If a model predicts so well that there is no added value from changes by human forecasters, will the model alone be sufficient? Given that meteorological systems are extraordinarily complex, humans will not be edged out in the near future, but perhaps eventually. As one modeler explained, "To predict something as complicated as the weather, you have to have something nearly as complicated" (Field notes). Tension is evident when a model is "wrong," when its modeled solutions are not predicting well.[73] Is the problem the scientific theory of the model or the data put in?[74]

Futurework on the Floor. Moving from the rarified climes of the quasi-academic setting of the Washington office to local forecast offices attitudes toward models become more complex. Given that I focus on the Chicago office, local skepticism may be greater, but attitudes seemed similar in Belvedere[75] and Flowerland. Theory threatens practitioners in that it situates expertise outside of the local office. It claims universality. Operational meteorologists, college graduates, had little experience in creating models, and were committed to their intuitive expertise in determining the forecast ("We know our weather"). The underlying concern is whether the forecasts generated by the models are sufficient or whether there is a visceral human understanding that cannot easily be modeled. The images of the future that the models provide—maps of forecast temperature or precipitation—look precise, but are these claims, generated by a set of equations without human input, sufficiently credible to be accepted without alteration?

For the extended forecast—particularly after the third day—forecasters often accept model predictions, but for the immediate forecast these men and women are skeptical of embracing models too completely. On one occasion in which several models were in close agreement, a forecaster noted with sarcasm, "At least in the *model world* there is some overlap" (Field notes). Distinguishing the model world from the real world was his responsibility. Another joked at a conference, provoking peals of laughter, "The real atmosphere has great difficulty simulating the modeled atmosphere, which has ruined a number of good forecasts" (Field notes). Or this dialogue:

> BYRON: Will there be rain tomorrow afternoon?
> EVAN: If you believe the Aviation [model], there will be rain tomorrow afternoon.
> BYRON: If you don't?
> EVAN: It will be tomorrow evening.
> MITCHELL: Or not at all. (Field notes)

Models depend on their data as well as on their theory, and sometimes these data are inadequate to generate reliable forecasts. If the model attempts to forecast for too small an area, it is likely to fail because of the lack of detailed data. As one meteorologist put it: "Models get better, but it makes the errors more extreme. They have gone beyond the point that their resolution [the degree of detail in the model] can handle. . . . The resolution makes it seem better than it is. It doesn't jive with reality. It's garbage, that's what it is. That's where experience comes in. To know whether it is real" (Field notes). On another occasion an office administrator told a forecaster that one of the models was going to change from a 12 km to a 5 km grid, meaning that the model would be forecasting for smaller geographical areas. He laughs, "You'll get to see a poor forecast up close and personal" (Field notes).

Even though models are relied on, forecasters warn each other about placing too much faith in them, transferring their authority to predict to distant academics. One meteorologist explained, "The modelers all believe that the models will take over the world. That snow event a few weeks ago. That's where we made our money. We are constantly thinking where we can add value to the models" (Field notes). Modelers admit that "snow" is their "four-letter word"; they forecast moisture less well than temperature. Models are not well equipped to deal with contingencies,[76] and moisture, so tied to local conditions, makes prediction difficult.

Any model can lead a forecaster astray. When I asked an intern about his greatest forecasting mistake, he describes an Aviation forecast. He forecast a low ceiling of clouds, "I had it for all the next day and the next day it was only scattered clouds . . . I believed the model data so much, you don't look at the reality enough. That's one of the big challenges, making sure that the reality goes with the models. If they're not jelling, you have to fit reality into the forecast" (Field notes).

Models are seductive in that they appear to be an end in themselves, an objective view of a real future rather than a hypothetical one. In their trappings of theory, they claim truth rather than guesswork. For former science undergraduates, such seeming precision is tempting, even if it limits autonomy. Because of its apparent simplicity and concreteness, the model excises some of the messiness of the real world. Further, since embracing a model saves time, they are even more alluring and perilous for the forecaster's authority. The model, as an organizing schema, provides an umbrella in a downpour of numbers.

The Beauty Contest. Whatever forecasters mutter, models are essential to knowing the future. This is certainly the case in the extended forecast, which is largely based on a forecaster's choice of a model, but it applies to current forecasts as well. No operational meteorologist would distribute a prediction without "interrogating" a model.[77]

The challenge for choosing a model begins with the recognition that models are not interchangeable. Even if they could be *objective*, they reflect differing assumptions about weather systems. One forecaster notes that models can produce dramatically different forecasts: "If one model brings a warm front in, it can be in the 60s. If the other doesn't, it can be in the 40s. That's a big difference" (Field notes). Indeed, once one model suggested a 25 percent probability of precipitation, while another claimed a 60 percent chance.

Models, like machines, become personified. They are felt to have *character*, and forecasters are told that their goal is "to put yourself in the model's shoes." When models don't match expectations, a forecaster may ask, "What do they know that we don't?" Model analysis has features common to psychoanalysis. Interpretation is not between self and object, but between selves, the mathematical one as mysterious as the one of skin and bones.

During the period that I was observing, the ETA model seemed to produce drier forecasts, pushing weather systems south; the Aviation model was wetter, pushing weather further north. Depending on which solution the meteorologist favors, they may personify the offending model: "Aviation is going bonkers on POP [probability of precipitation]" or "That ETA looks as if it's been out to lunch" (Field notes). This can be extended to the claim that models will run "in streaks," in part a function of seasonal equations. One forecaster noted, "I don't think there is a preferred solution. Everything I've heard says they tend to run in streaks, where one does well for a while, and another one does better for awhile longer" (Interview).[78] As Elliot Abrams, Chief Forecaster of the private firm AccuWeather, remarked of a successful model, "Who am I to say the numerical guidance is wrong[79] . . . When a model has a hot hand you stay with it."[80] The belief that models can have "hot hands" has led the National Center for Environmental Prediction to establish on the fourth floor (the data analysis department, not the modelers) a "model diagnostic desk" that examines the "biases" of models and shares the biases with local offices. As one of these analysts explained, "We have a feel of when the model is overdoing it or underdoing it . . . . We're suspicious. We have to let the forecast offices know when the models are not perfect" (Field notes). This attitude provides some interpretive control over a process that may be perceived as having a life of its own with its streaks and idiosyncracies.

If models are treated as errant children, not simply bundles of equations, which child does one favor? Or using a different metaphor, "It's like playing golf. Everything sort of depends on the situation. What kind of club you will use" (Field notes). Despite their divergences, each model relies on the same information. The data are identical, and so the differences are based in theories about atmospheric behavior. To be sure, differences are often a matter of degree. As Randy notes, one inch of snow might be 3 inches, but it is rare that partly cloudy becomes an inch of snow. Given this, forecasters search for clues that models are converging. As one forecaster informs her replacement: "The models are all over the place. It's lovely [said sarcastically]. ETA is in and out quickly [with little precipitation]. The Aviation is wet. It's phased slow. I tried to go with a compromise" (Field notes). The interrogation of models can be lengthy:

> When I arrive Sean tells me, "You've come on a good day. A lot of things to consider. I have a lot of tough decisions to make." He is attempting to select a model as the basis of his forecast.

> Severe weather may be coming, and the models make different predictions. Throughout the shift Sean compares models. He tells me: "Personally I'm favoring the Aviation. It's quite possible we'll have some thunderstorms tomorrow. . . . It doesn't look as if it's going to be gigantic. The models are all over the place." Later Sean tells me that he thinks that there must be a problem with the initialization of the Aviation model. "The Aviation starts with bad data. Maybe the model was running off the wrong assumptions." Sean says to Patrick, "If you believe the ETA, something could be knocking on our door at 18Z [Zulu or Greenwich Mean Time or 1:00 p.m. in Chicago]. I think that's too fast. . . . I'm going to go 50 [percent chance of precipitation] for tonight. Randy [the evening forecaster] can sort it out." (Field notes)

By the end of his shift, he has moved toward a forecast based on the ETA model. What is critical, however, is that he is attempting to shape his prediction in light of these two alternate realities. In generating predictions, forecasters are tethered to those authenticated claims that are available.

Running against the Models. The threat for operational meteorology is that models will someday overtake them; experience will not match theory. Operational expertise is placed against the expertise of the theorists. Modelers encourage this anxiety, even while they formally deny that they have any such intentions. This fear, veiled in velvet words, pervades the insistent striving by operational meteorologists to demonstrate that their work has a value-added quality.

The fear is simply put. If models can predict weather conditions as well as local forecasters, what need is there for these superfluous employees? Job security is tied to issues of occupational self-esteem. Perhaps local forecasters are essential for the occasional severe weather outbreak, but some suggest that even this service could be provided by regional offices. Defensiveness is evident. Forecasters are like professional chess players threatened by computer scientists:

> [Models] are not gospel. The scientist can add value to what the models have, especially in the short term. . . . We have performance measures that we can show that we consistently show improvement over the model guidance. (Interview)

: : :

> [You need to] find little nuances that you could tweak and make it better. If [meteorologists] start forecasting everything that's coming across in the models and not actually putting their own personal interpretation to it, they're going to flat out sink. You're doing no better than the models, why are we paying you? (Interview)

These are tough issues, and ones that will become more salient as theory continues to improve, tempered by what works. How far can this streamlining go before only a single forecaster remains in a high castle outside of Washington?

As a result, forecasters compete against models. As one forecaster explained, "You want to beat the model. You want to walk away feeling that you improved on the model, and you don't want to come back the next day and find out that the model got it within two degrees and you were four degrees off. If that's what happens consistently, there's no reason for being here" (Interview). The models are spoken of as "the competition" or "what we are trying to beat" (Field notes). The system is structured so that this competition is inherent in the process of forecasting.

The emotional strain between human and model is particularly evident in the struggles of young forecasters. As one lead forecaster reminisced: "I think most [forecasters] want to do what's right. We're appeasing the weather god, not the numbers . . . [As a young forecaster] I was afraid to vary too much from guidance. I was looking at what guidance was telling. I was young. I was told to look at the numbers. Forget the guidance! Just put out what you think is going to happen" (Field notes).

For a novice lacking confidence in pattern recognition, this autonomy may be easier said than done. Older forecasters explain that when they compose their forecasts they avoid examining guidance too closely, feeling that this constrains them. One remarked about the focus on models, "You become conservative. The models can be your worst enemies. You're always trying to beat the models" (Field notes). Heroic narratives are shared about forecasters who had a "touch" or a "sixth sense." These forecasters were said to have an intuitive grasp of the changes of temperature or precipitation and could beat guidance. The models are seen to distract from the forecaster's authentic knowledge of how weather operates. Theory comes to stands apart from practice.

Autonomy in Podunk. Models, apparently definitive in their predictions, lead forecasters to worry that they are—or will become—typists or machine tenders, tidying up after an autonomous model. This view is exemplified in the assignment of prestige to modelers. Forecasters have been trained to follow the long-distance instructions that they have been given, limiting autonomy; they consume documents but are not capable of producing alternatives.[81] Their autonomy is highly localized. In contrast, modelers are credentialed scientists. While forecasters typically have B.S. degrees, most modelers are Ph.D.s., higher in the scientific pecking order. A sympathetic modeler referred to the unkind remarks of modelers who sometimes treat forecasters who criticize their work as do mechanics who must deal with customers who ignorantly question their repairs.

Modelers may be given deference but not always with good grace. One modeler, whose position requires him to interact with forecasters, distinguishes between Mecca and Podunk, at least in the minds of modelers and in the resentments of forecasters. He notes that "some [modelers] treat [forecasters] as useless pieces of machinery" (Field notes). Forecasters have their own images: "I don't like to say they're the evil empire. . . . Their concept of how things work out in the field is warped. . . . They are divorced from the field. . . . There's not a real good communication, not much of an interaction with people in the field" (Field notes). This meteorologist notes that there is a program whereby forecasters can visit NCEP offices in Washington, but no similar program in which the modelers trek into the field.

When guidance is newly available forecasters must decide whether to change their predictions, a choice that can reveal autonomy or deference:

> Before Sid sends out his forecast, he waits for the Meso-ETA guidance, which typically is available at about 3:00 p.m. He tells me that a few days ago, "I had the whole thing typed up, but I made some last minute changes. The model was tending in a particular direction. I'm glad I did, because later models showed the changes even more dramatically. But my philosophy is not to put too much in. *These models go off on their own.*" (Field notes)
>
> : : :
>
> At about 3:15 p.m., a new model is available, calling for less rain than the previous version. Cal, the forecaster on duty, is upset, and says, bowing to the model: "I guess I have to compromise and go for a *chance* of rain. Try for some sort of compromise. We may get grazed. I'll back off. . . . It's frustrating. You think

> you're putting out crap. You take a swing at it. It's out. It's not very good. I just said chance of rain. It's the best I can do." (Field notes)

These forecasters feel constrained by the autonomy of theory, even though in principle they have the authority to go with their gut. The question is how far do they feel that they can stray from the model, given organizational preferences that require that any major deviation from guidance be coordinated with neighboring offices.[82] In the words of another forecaster, one must avoid "hugging the model," but use it to direct one's thoughts. Autonomy for a human forecaster resides in the ability to select which model they choose to rely upon—on which mechanical bull will they ride. Acting autonomously demonstrates that they are professionals with some measure of predictive control, even if they remain within organizational expectations. One forecaster claims that he has about a 90/10 reliance on guidance, a function of the increasing sophistication of models, and the dangers of deviating too far, but he adds, "Now it's really come to the point, because there are so many models out there, is the ability to pick which model is the correct model . . . because you're going to be offered a multitude of choices" (Interview). By being offered choices, the forecaster's autonomy can be preserved, even when the issue becomes which theory is best. The local office is presented with a choice of plausible futures, and within this multiple choice, they choose the one they think is best.

Historicized Experience

The third element of futurework involves incorporating the experience of the forecasting community into the practice of prediction. Data are assessed as claims about the reality of the world. Theory, as embedded in meteorological models, makes claims about how this present should be extrapolated into the future. In contrast, subcultural knowledge incorporates tacit, local knowledge. It valorizes the authenticity of occupational experience. Competent forecasters know what they need to recall about the past and, implicitly, what they can forget.

The social domain of knowledge challenges the apparent objectivity of technology. Through claims that it incorporates both theory and data, technology is posed as the core of forecasting.[83] Technology can come to represent what Erving Goffman speaks of as a *primary framework*,[84] a view of the world *as it really is*, rather than a transformation (or *keying*) of that reality, according to a set of theoretical procedures.[85]

Technology produces expertise but undercuts authenticity. While meteorologists point to the intuitive wisdom of farmers (institutionalized in *The Farmer's Almanac*), they are also implicitly justifying their own experience as long-time observers of the skies and as a community that shares that knowledge.

One forecaster speaks of *meteorological cancer*, noting that "people can rely too much on technology and remove themselves from looking out the window. Sometimes people will forecast things and they can't tell you why. Sometimes you can't even explain it. It's pattern recognition. It's experience" (Field notes). This rhetoric challenges alienation and deprofessionalization, justifying the position of operational forecasters.

Several meteorologists claimed that one can have "too much technology," allegedly producing less accurate predictions and providing cognitive overload. The present can be too much with us. As one forecaster mused, "The problem with modernization is that there is too much to look at. We're overwhelmed. Our rule of thumb is when you can't see the map [because of overlaid bands of data], it's a bad sign. So our goal is to figure out what we need to look at on a smaller scale" (Field notes), He speaks of a process of triage of technological inscriptions.

However, forecasting doesn't just depend on personal beliefs but on group interpretations. The social enriches professional memory. When severe weather is deemed likely, dialogue is particularly common. Forecasters attempt to interpret present conditions, how the conditions link to theoretical models, and how they relate to previous cases that reflect a similar pattern. During my observation a storm near St. Louis was developing by a synoptic low, "a nice place for a storm to be." A colleague reminds him that was the case for the infamous April 1925 "Tri-State Tornado" that ripped through Missouri, Illinois, and Indiana, the deadliest tornado in American history.[86] These forecasters are collectively testing ideas of this storm's future with their collective memory of storm activity. They historicize the present, attempting to reduce the uncertainty of the future by fitting it to the past. Unlike data and theory whose social origins are elided, collective experience is an *explicitly* social phenomena. If it lacks precision, because it is social it asserts an authenticity that other claims cannot match.

Operational forecasters often admit a lack of certainty ("The jury is still out," "We don't have a sure idea of what will happen," "There is too much uncertainty"). While such rhetoric might seem to undercut their authority, in practice it does the opposite, suggesting that general rules of interpretation are of less use than the local knowledge that comes from the analysis of storm data within a community of experience. In these cases, forecasters debate and discuss the developing weather in internal dialogues (made audible through the presence of an observer) and with their colleagues. One forecaster struggles while simultaneously embracing his authority:

> The environment would certainly favor the development of large and serious tornadoes.... This front would have to race up. If it does, all Hell could break loose. The predictability is very low. Most of them don't work out, but the ones that do, kill people.... It's going to be challenging to decide whether to issue a box. The question is whether there is enough hail to warrant a watch.... It is really complex to know whether the storms will be severe enough to issue a watch. What do you guys think?... It's really possible. There's juicy air over the top of that. My guess is that it's going to be a late show. I would not expect it to be before ten o'clock.... It's not clear if it will be anything more than a squall line. Many, many questions. (Field notes)

Choices must be made, privileging the worker as seer over data and theory.

Progress often occurs without workers' being able to specify its theoretical or empirical grounds. Much knowledge is intuitive or tacit.[87] Of course, we do construct *post hoc* explanations as necessary. The processes of work often rely upon a taken-for-granted reality[88]—an absence of formal rules of interpretation, following the path laid out by philosopher Ludwig Wittgenstein.[89] The reasoning skills of scientists and other technical workers share much with others in mundane contexts.[90]

That this applies to operational meteorology has long been recognized. In 1917 meteorologist George Bliss explained: "The most profound students of atmospheric physics... were far from being the best forecasters.... In order to excel in the profession one must possess a special faculty for intuitively and quickly weighing the forces indicated on the weather map and calculating the resultant."[91] Maja-Lisa Perby finds the same phenomenon among Swedish forecasters, suggesting that those with a "comprehensive idea" of the weather will do better than those who rely on computer systems alone.[92]

One forecaster in emphasizing that he believes the best forecasters are gamblers and poker players, explained, "Sometimes people will forecast things and they can't tell you why. Sometimes you can't even explain it. It's pattern recognition. It's experience. It takes a real skill, a real feel, a hell of a lot of expertise. You'll never see the same thing twice. That's the problem" (Field notes). Another forecaster remarked: "Your gut feeling would now tell you, I've seen this before. I know it's going to go like that. You just know it's going to pop. [He was forecasting heavy fog while in the Air Force]. There was a lot of tension to get those flights off. Why is this rookie trying to tube our missions? It was as clear as a bell, and I just turned around after a briefing, and it was just like pea soup. . . . We had misses, too, but this was one of our glories" (Field notes).

After one of those misses, a forecaster explained, "The atmosphere was trying to tell me one thing, and I wasn't listening" (Field notes). In a similar vein I was told that you could feel, smell, and even hear weather: crisp weather sounds clearer, moisture filled air sounds duller and has a "musky smell." Another describes his awareness of weather as constituting a sixth sense.

This process requires adjusting to local knowledge.[93] "Each office has its local weather and each office has its local hazard." As noted, when forecasters transfer, as they often do, their intuitions must shift. One commented, "I'm from North Platte. I got used to the way weather was down there. I have to adjust to this weather" (Field notes). The claim that weather knowledge is only grounded on a formal evaluation of data and theory is belied by the intuitive understandings of those who forecast on a routine basis. Even though these workers do not have primacy in access to data or theory, their personalized knowledge makes their knowledge authentic in ways that formally created data cannot be. As is true of expertise generally, lived experience plays a crucial role in the ability to perform specialized tasks.[94]

Informal knowledge has its costs, however; if in no other way, it is hard to justify rhetorically both inside and outside the occupation unless there are some techniques that can be pointed to. Although becoming a professional requires formal training, it also requires becoming socialized to communal work practices. These practices constitute *rules of thumb* (or finesse rules), practices that are known within the boundary of workers. They belong to the oral tradition of work worlds.[95] These strategies stand outside academic theory, but they work well for forecasting, providing an alternative basis of expertise. These rules make occupational uncertainty routine, much like the implicit timetables by which physicians organize the prognosis of patients with chronic illness.[96] Rules of thumb regularly appear in meteorological forecasting:

> Randy explains the difficulty in forecasting the heaviest band of snow. He says that to find the snow track, often a narrow band, he draws a line 200 miles north of the low. In predicting

> the path of the low, forecasters feel that they can predict heavy snow. (Field notes)
>
> : : :
>
> One of the rules of thumb for predicting severe weather involves temperature and dew point. Gordon tells me, "You like to see a ten degree spread between temperature and dew point." (Field notes)
>
> : : :
>
> Davis asserted that adding ten degrees to the 10:00 a.m. temperature to get the maximum daily temperature "works more often than not" in summer (Field notes)

These examples could be multiplied, as there are strategies for determining the likelihood of fog, cloud cover, and so forth. These knowledge strategies constitute subcultural features of scientific work, valued by practitioners, but separate from formalized knowledge.

A distinction exists between "hypotheses" (the rhetorical tool of academics to specify relationships among variables) and "rules of thumb" (the rhetorical trope used by practitioners). The latter were not formally sanctioned but were treated as *tricks of the trade*. When I inquired about how such techniques were learned, one forecaster responded: "Some of it gets spread around. . . . It's one thing to tell you the rule, and it's another thing to see it written down. It's *finesse rules*. It's based on science, but it's also based on experience" (Field notes). Perhaps they are not written down in that they lack a theoretical infrastructure, having a "seat-of-the-pants" justification. Of course, in practice, some academic research also lacks a strong theoretical base but gains legitimacy from its institutional origins, and thus an implicit theoretical infrastructure is assumed.

These techniques reflect occupational make-dos, part of the subculture of operational meteorology. In contrast to his reliance on rules of thumb, one employee explained that he mistrusted those who work from hypotheses, which, in his view, "introduce bias" (Field notes). While his comment is extreme, workers often create their forecasts intuitively and then justify them in a *post hoc* fashion, using theory as a rhetorical resource. For instance, after one particularly complex forecast, the meteorologist on duty turned to his colleague and remarked in reference to the area forecast discussion that justifies the forecast, "Now I have to come up with an AFD, so I'll have to explain it somehow." His colleague responds, "You'll dazzle them with science" (Field notes).

While forecasters do rely heavily on academics for their theoretical models, just as they depend on engineers and technicians for their data, constructing a future is a form of collective action.[97] Operational meteorologists create their own forecasting solutions on the shopfloor, part of what Karin Knorr-Cetina has termed their *epistemic culture*.[98] This culture cedes authority to those who work the floor, the location where "real" forecasting gets done.

The Practice of Futurework

In preparing a forecast, data, theory, and experience mix. Granted, for routine diagnoses—meteorological, medical, or otherwise—a professional relies upon standard procedures. But at times things get dicey, and others are invited into the conversation. In these conditions of uncertainty, it is through the power of collective assessment that the future is hammered out. Forecasters continually test ideas and review uncertainties in light of the data at hand, the rules of extrapolation, and their experience.

When colleagues are present, they too may be invited into the dialogue. To appreciate how futurework gets done as a collective project I focus on a sunny 50 degree day in March in which the challenge is to determine the snowfall for the next two days. For four hours I observe Randy as the public forecaster of the day. During this period he talks with several colleagues in the office as well as with three other offices. On the midnight shift Evan had forecast "significant accumulation possible," but he did not issue a winter weather watch or an advisory, although other offices had issued watches. Now the question for Randy was what to do about the approaching snow band. How much snow should he forecast and is the amount sufficient to issue a winter storm watch:

> Randy reviews the available models, the maps of ground and upper air observations, and the radar images and the satellite pictures. He comments: "The new models all have the snow further south. There's a stream coming out of the south. That may limit moisture to the northern stream. That's tricky. . . . [The models] are preferring the further south solution. [Sid: "If they favor the ETA that keeps the threat out."] The ETA tends to open up a little faster. There's a lot of dynamics to work with. . . . NGM follows the Aviation [a model that predicts substantial snow over Chicago]. The ETA of late hasn't performed the best, but they're all flopping around a lot. . . . I don't know about the heavy rain

> tonight. [Byron: "How does it look? Better or worse?"] If you buy the ETA, we're up at the edge of the precipitation. If you buy the Aviation, we're in the middle of the precipitation. The models are pretty similar, but we're at the point that a little difference makes a big difference. You see with the ETA, we're really on the edge of things. [Byron: "It is a little odd. It's not something we usually see."] The question is what to do. [Byron joking: "You have eight hours."] In the first twenty-four hours the models are very similar. [Byron: "The difference is between a rainy day and something more significant."] Even if you go with the ETA forecast, you still end up with accumulation. . . . I'll have to shift where [Evan] has his significant accumulation. I hate to put that out. [Byron: "Yeah, the yo-yo effect." . . . Sid: "Decisions. Decisions. Decisions. If it were me, I wouldn't put [a winter storm advisory] out. Right now it would be very debatable if we get 6 inches [the criteria for a warning]. To me the biggest fact is that we get this coming in with no moisture source."] My inclination is just to hold off. Mention accumulation. There are things about this that don't seem right. What concerns me is that this first wave takes all the moisture out. I hate to go for a watch. While we may get several inches, it isn't clear that we will meet warning criteria. I'm sort of leaning to the ETA. I inherited a forecast that has significant accumulation up through Thursday night. The other thing I've inherited is heavy rain. I'm not too sure about that. It may develop late. [Martha, an HMT: "Are you going to bite the bullet and put out a watch."] I'd have no good reason for doing that. There's too much uncertainty. [Martha: "We're scaring all the public. They say there's going to be 6–10 inches of snow."] The potential is there. We could see a few inches—3–5 inches. We're not completely sold on the idea that it will be a heavy snow. I'm going to be real conservative. It doesn't look as if it's going to be as bad as it could. I'm just going with accumulation possible. I'm not going to be cute. Maybe there will be an advisory. [Sid: "How are you going to do this with all your moisture taken out."] I'm not going to take it out, but I'm going to downplay it. Going round and round in a circle. Things can change at any moment at the whim of the forecaster. (Field notes)

This lengthy excerpt demonstrates how forecasts become collective products. Claims are constructed from an abundance of evidence: data, theory, and experience. They rely on improvisation that builds on the rights of the professional to know. Randy tries to make everything fit into a seamless package, a challenging task, particularly as the data keep changing, but one for which he can gain advice and sympathy from his co-workers.

Ultimately both he and we must recall that none of the evidence reflects what *will happen*, but only what *has happened* and what models suggest *might happen*. Several options are possible; certainty isn't, until after the fact.[99] To gain confidence, one hopes to create a forecast community, building on the input of others.

On this occasion Randy's prediction was generally accurate, but on other occasions a similar process led to blown forecasts. Collective decisions are not always correct ones, but they are necessary. In predicting the future, Randy takes the observations that are available, attempts to integrate them with the models he has been provided and, in doing so, mediates the data and theory through his experience and his gut response. He relies on colleagues, attempting to fit their interpretations into his own, and providing a *community of expertise*. By the end, Randy is still unsure, but he has persuaded himself of his own forecast answer.

Building Legitimation

Forecasts demand audiences, and these clients judge the plausibility of the forecast and the credibility of the forecaster. Legitimation produces the forecaster's right to speak. Every act of prediction is set within an institutional nexus. In chapter 6, I discuss in greater detail the relationship of forecasters to their primary audiences—the public, private industry, and the media. Here I argue that it is the reality that government forecasters are socially sanctioned sources of meteorological information that gives them authority to make predictions on which citizens rely.

Forecasts are a routinized product of meteorological organizations. Every day before dawn (4:00 a.m.) and during mid-afternoon (3:30 p.m.) the Chicago office issues its forecast for the next week (with separate day and night forecasts for the first three days[100]), available for morning and evening newscasts. The timing of forecasts is linked to the requirements of other institutional actors within the organizational field.

Legitimacy involves both creating a positive reputation and avoiding a negative one, as well as creating links to others that publicize one's credibility. If the public and other organizations are not concerned about forecasting errors (missed temperatures, amount of rainfall), credibility can be preserved. Being correct is desirable, but it is preferable not to be *wrong*. As one forecaster remarked, "When the rubber meets the road, we are paid not to be wrong. We're the no surprise weather service." Said another in the same vein, "I don't think we're paid to be right" (Field notes). Being wrong involves the public recognition of error, so one can predict wrongly without being wrong if the public doesn't know or care. Media and private weather firms wish to avoid public complaint and the public and private industry desire not to be inconvenienced. If a forecast is inaccurate but not inconvenient, it is as good as being correct. The integrity of the weather service is preserved.

As a result of organizational demands, caution is built into forecasting. To avoid being wrong, forecasters strive for predictions in which error will not be noticed, as opposed to producing the forecast that represents the meteorologist's best judgment. If rain is possible but not likely, a forecaster may include a chance of rain to avoid blame if the day is damp. On a day in which snow was expected, I observed the public forecaster struggling to determine what he should forecast as the likely snow amount, a difficult decision because bands of snow can be narrow and their speed uncertain. He expects snow, but he is also concerned about how his forecast will appear in light of the storm's aftermath. He explains, recognizing an organizational imperative: "We can't say accumulations are possible. We have to say how much." In coordinating with another office he remarks: "I was thinking of going in the 1–3 inch range, and maybe a little less near the lake. That way if they get a slushy half inch, [the forecast] won't be so bad; if they get 4 inches it won't be so bad."

The forecaster is forced to predict the future, not in its own terms, independent from the social world, but as a form of impression management. The prediction, even when wrong, must reveal sound professional practice. Although weather forecasters do not have the threat of malpractice to motivate vagueness, they do have political audiences that hold the organizational purse strings. This reality encourages strategies that permit claims that are adequately correct, revealing the agency as bureaucratically competent. Fortunately for meteorologists, the indeterminacy of forecasts is assumed, providing a legal barrier against accusations of malfeasance. Courts have held that erroneous weather forecasts, even those leading to loss of life, are not negligent in that they are: "a classic example of a prediction of indeterminate reliability, and a place peculiarly open to debatable decisions."[101]

Despite their legal shielding as a function of uncertainty, forecasts are embraced by audiences. This legitimacy operates along several dimensions. Forecasts can be defended because of the standing of the occupation, the organization, or the forecaster, relying on justifications from professional practice, institutional power, or self-presentation. Each involves a form of impression management but with different bases of authority.

Occupational Legitimation. The public holds images of occupations. It is not only that we evaluate individual workers, but we evaluate them as representatives of an occupational class. While the public is familiar with particular media forecasters—and may have positive or negative assessments—those who stand behind these celebrities are anonymous. They are seen simply as "weather forecasters" or "meteorologists," but not as individuals. Legitimation applies to the occupation as a whole, rather than to individual practitioners. Forecasters are interchangeable, except as they appear as media performers. The public belief that forecasting represents scientific practice supports legitimation, and meteorologists do what they can to persuade the public of this fact, nowhere more than on media venues, such as the Weather Channel with its "Expert's Desk," staffed by a Ph.D.

When one examines occupations engaging in futurework, the degree to which an individual persona is tied to predictions varies. Doctors, for instance, are linked to forecasts in a way that does not apply so tightly to economic planners. In the case of meteorology the media broadcaster reaps the tributes and insults of the public, while those who create the government forecast on which most media broadcasts depend remain anonymous.

Because of the importance of occupational credentialing, meteorologists desire to justify their work to the public. This process is evident in the certification procedures of meteorological organizations in awarding credentialed broadcasters a professional "seal of approval." Displaying this certification can be used by a broadcast station in their advertisements. The desire on the part of the National Weather Service to hire forecasters with college degrees in meteorology has a similar role because a degree vouches for the occupational integrity of the novice forecaster.

Organizational Legitimation. The meteorological forecasters examined in this study are employed by the National Weather Service, an agency within the United States Department of Commerce. Their predictive adequacy is, to some degree, enmeshed in politics. They represent not only themselves and their colleagues but the government. The agency establishes a series of steps through which forecasters must proceed for promotion. The novice progresses from intern, lacking the authority to distribute a forecast without approval and required to pass a series of organizationally mandated tests, to journeyman (or general) forecaster, and eventually to lead (or senior) forecaster. Although the lines of authority between the last two positions have blurred, the promotion reveals that the organization has judged competence. At the time of this research, weather service offices typically employed five lead forecasters and five journeyman forecasters, and so, unlike university tenure decisions, promotion requires a vacancy created by retirement or transfer.

It is not only meteorologists who justify their predictive skills but their institutional handlers. The accuracy of forecasters advertise government competence, and so bureaucratic leaders place the best possible face on their work products. The successes of the agency are trumpeted to politicians and the public with failures hidden, whenever possible. The Plainfield disaster, as described in chapter 2, delegitimated the credibility of meteorological forecasts, just as the success of forecasting tornadoes raises the agency's profile. Given that few members of the public keep track of how accurate the government forecasts are, the claims are presented by interested parties: the agency or their media critics. In each case, however, what is being presented is a claim of organizational competency. When the claims are favorable, it is the organization that is legitimated, not particular employees.

Presentational Legitimation. Legitimation also depends on how individual meteorologists present their predictions. A proper presentation provides authority through the form of the forecast. The claims must be presented with confidence and clarity. Given uncertainty, this is not easy, but, to be a competent professional, the forecaster must pretend that certainty prevails. There is little room for waffling, outside of the legitimate options of providing ranges of temperatures or likelihood of precipitation. "Don't know" is not an option. A meteorologist points out, echoing Erving Goffman's[102] concept of dramaturgical circumspection, "You've got to put something out, but often we don't have a lot of confidence in it. Maybe 20 percent. You have to put a face on" (Field notes). Audiences demand a forecast, presented with confidence. While the forecast can, of course, be changed, the prediction is to be accepted until such time as it is replaced.

The public and the media rely upon these presentations, implicitly accepting that a bad forecast is preferable to no forecast. It is the very fact of the current forecast and the routine sequence of forecasts, phrased in a formulaic way, that demonstrate that meteorology is legitimate. Public confidence is as powerful a justification for belief as is the accuracy of the claims.

Futurework

To prognosticate is dangerous. One puts oneself on the line. What is happening is an uncertain gauge as to what may continue. Digging into the past to understand the future is to peer into smoke and mirrors. Yet, this does not suggest that forecasts are produced without considerable ability. They are often correct, at least in the public judgment of such things. Occupations are given credit for their ability to announce how the future will unfold. These occupations depend upon the combination of data, theory, and communal experience. But to receive public confidence their claims must be institutionalized and culturally endorsed.

Weather forecasting is impossible without the vast data provided by technology. Operational meteorologists work in spaces stocked with machines. These machines do not just exist but exist because some person, group, or organization believes that gathering data aids in understanding the present and the future. Someone believes that accounts of the past are needed to access the future. Machines are the means by which theories give birth to theory, through the midwifery of data and data handlers. The presence of machines provides the magic and the justification by which predictions are made and accepted.

The justification for predictions also derives from the authority of the model with its powerful, yet implicit, claims of expertise. Models too rely on interpretations of the past as they make claims of the future, and the fact that models disagree mean that they are interpreting the past in different ways as they present their predictions. In this domain, operational meteorologists set themselves up as judges, making a choice when models conflict. The forecaster is the honest broker, relying on communal knowledge. While models have improved in their accuracy, there are now a wide array of available models among which to select, a process that sometimes involves personifying the models.

Judging the meteorological future depends on a selective interpretation of the present. The future is largely a function of a continuation of the *conditions* of the present, coupled with a continuation of the *trends* of the past, a stance that is as true for financial markets, bodily ailments, psychic distress, as for the skies. If changes in direction are proposed, a theory must explain why equilibrium and current trends are to be ignored.

Meteorologists rely on a set of knowledge claims: part experience, part intuition, part subcultural wisdom, and part scientific claims. What is to be remembered is selected based on which memories are judged relevant to the work at hand. These claims are filtered through the assurance of the meteorologist in the hot seat: the individual who is given the authority and the responsibility to know. The question becomes, given these domains of knowledge, how does the operational forecaster demonstrate expertise.

Just as the future is socially organized, so is the act of predicting that future. Futurework results from the linkage of occupational skills with organizational resources. Occupations have rules by which they gauge what is likely to happen—they have modes of extrapolation, theories, and experiences that permit them to guide that extrapolation. But more than this they rely on an institution that supports their claims. Even if their public does not accept these claims in their totality, they do accept the possibility of prediction.

While meteorology reveals how close futures are established, futurework is not limited to this domain. We rely on doctors to make forecasts. How are they able to do so? Part of their skill derives from the bodily data to which they have access. Added to this is the organization of scientific knowledge into a theory of disease, tied to the experiences of the doctor or her peers. Finally doctors claim for themselves and are given a respectful cloak of expertise. We assume that they can forecast the future. In part this results from the life experience of their clients with other doctors and part is an acceptance of the claims made by their institutions (hospitals, medical societies).

Financial planners have similar challenges in claiming an uncertain future. They have extensive information about the economy from which they can draw, but what do these measures presage about how the economy will respond in the days and years ahead. As with meteorologists and internists, clients depend upon their expertise in making financial choices. The financial analyst has data from which extrapolations can flow, theories that are both general (predicted rates of growth) and local (conditions that might modify these rates). Finally, these analysts are employed by organizations with more or less public esteem and confidence.

An extreme case is the fortune-teller. Fortune-tellers differ from other predictive occupations in several ways and in this serve as an archetypal case of the problems of prediction when one lacks widespread public confidence, although their clients, presumably, find the making of such predictions plausible. The fortune-teller is limited in the knowledge available. The art is to infer from client's cues information that is otherwise hidden. Gaining data for extrapolation is a challenge. From these scraps of data, coupled with folk theories of human nature, the fortune-teller creates an explanation that clients may accept. Different fortune-telling traditions have distinct theories of the future—psychological, social, religious, or mystical—each leading to a set of confidently presented claims (whether or not the fortune-teller believes them).

Predicting the future is social, both in the act of prediction and in the organizational infrastructure that permits its acceptance. As an occupation with the responsibility to peer into the future, the future of weather forecasting itself, constrained by the development of expert systems outside of the control of the forecaster and judged by potentially critical audiences, is uncertain. The public demands to know what the heavens will bring, but at what cost and on whose authority remains cloudy.

4 Writing on the Winds

Probable northeast to southwest winds, varying to the southward and westward and eastward, and points between, high and low barometer swapping around from place to place, probable areas of rain, snow, hail, and drought, succeeded or preceded by earthquakes, with thunder and lightning. **Mark Twain**[1]

Weather is a literary speciality, and no untrained hand can turn out a good article on it. **Mark Twain, *The American Claimant***

: : :

Operational meteorology is drenched in discourse. Forecasters write, and, as much as the examination of maps and numbers, this literary activity defines who they are and what they do. For much of the history of the National Weather Service the job of the forecaster was as much written communication as scientific analysis. Indeed, the two were inseparable. Scientific analysis had no value without the ability to share it. Admittedly the writing was formulaic, but sometimes formulae focus greater attention on linguistic choices. By speaking to a public audience the forecaster gained authority. For this realm of public science—and implicitly other occupational domains—the meaning of work was established through the practice of writing.

In this chapter, I explore public science as communication. Specifically I address four aspects of the occupational tasks of meteorologists: (1) how they coordinate their forecasts with others inside their office and with other National Weather Service offices, (2) the art of writing forecasts and forecast discussions, suggesting how meteorologists think about their words, (3) how forecasters at the Storm Prediction Center use visual representations ("boxes") to claim their authority, emphasizing that communication is not necessarily tied to words, and, (4) the technological change that I observed during my research in which a computerized forecast system was introduced. In this system meteorologists manipulated a database, which removed the authority to create the written forecast from the meteorologist.

Inheriting the Wind

Meteorology, like most professions, depends on coordination. Even if forecasters frequently stare at their screens, jotting notes, drawing or typing, at times they emerge from their technological cocoon and ask for advice.[2] The staff routinely discusses current weather conditions, even when the most introverted or focused forecasters are on duty. While questions are often directed from interns to forecasters or from journeyman forecasters to lead forecasters, following status lines, comments can be directed in any direction. Of course, when they proceed from junior to senior the remarks may include a dollop of deference. Forecasters may even stroll outside together to touch the air and interrogate the sky, merging the cognitive, emotional, and social.[3] Talking about the weather is satisfying; it is what these "weather weenies" most enjoy. While the collective creation of weather forecasts is most evident under threat of severe weather,[4] collaboration occurs on less dramatic occasions as well.

Two rituals establish connections among forecasters: *briefings* and *inheritance*. When a forecaster arrives for duty his predecessor provides a synopsis of current and future weather conditions. The departing forecaster sits at the terminals, as the arriving forecaster stands behind the chair, showing deference to the *authority of the hot seat*. One meteorologist made this explicit, explaining when his colleague exits the area for a moment, "I won't sit down until Evan leaves" (Field notes). The briefing is usually succinct, no more than five minutes. The departing forecaster describes the complicating aspects of the forecast, what might change or demand attention in the next hours, or what equipment is not functioning properly. Staff have different styles and preferences; some favor broad accounts, which permit them to explore independently, while others prefer more detail. On days when a potential for severe weather exists, the briefing is more extensive, and the forecaster from the previous shift will not depart until his or her replacement feels comfortable. As noted in chapter 1, forecasters use Lysol to remove the physical traces of their colleagues, yet they cling to the presence of their predecessor's work via the continuity of the forecast and expectations for the future.

This routine of sequential communication links the day shift, the evening shift, and the midnight shift. On days on which severe weather is unlikely, the evening shift is transitional; it updates conditions and is present "just in case." As a result, the evening forecaster doesn't need the detailed knowledge of weather conditions that other shifts require. The evening shift is a placeholder for those shifts—day and midnight—that are responsible for issuing forecasts. The evening shift focuses on what must be communicated to the forecaster on the midnight shift. As one Flowerland forecaster remarked, "What I want to hear is what is unusual and what is weird, and if I'm the [evening] shift, what I would want to push on to the midnight shift." The briefing directs the forecasters to what to look for and may direct them away from other aspects of the weather.

Once the previous forecaster exits, the person newly on duty *owns* the forecast. He or she has *inherited* it. In theory, it can be changed as much as desired, although in practice norms exist for what gets altered and when. Forecasts are shared between those whose shifts abut, and this means that the possibility for ill will exists if revisions are too rapid or extensive or are not based on major meteorological changes. One's forecast is an extension of one's self. In this case inertia may contribute to harmonious social relations.

Although a meteorologist owns the forecast, constraints limit those modifications that are considered legitimate. The forecast is owned by the office as well. Meteorologists are to avoid "flip-flop," "yo-yo," or "ping-pong" forecasts. These terms do not refer to changing a forecast, but rather to changing a forecast back to what it had been in a previous iteration. The forecaster whose words and ideas have been altered has a crucial choice. Should he return to the previous forecast? Until there is new evidence or others are persuaded, a return to a previous forecast can discredit a stubborn forecaster unless a forceful account is presented for the change.

Forecasters routinely work the same shifts for a week. As a result, if the day forecaster and the midnight forecaster disagree on weather conditions, the forecast could shuttle back and forth between the predictions of the day worker and those of the midnight worker, confusing the public and the media. Social control is required:

> I've got this philosophy, you don't change the forecast unless you really have a significant difference in the previous guy's forecast. You ride with it. I don't want ping-pong forecasts. You better be darn sure that things are going to change, or else you stay with that previous guy's forecast. And I tell the previous guy that makes the forecast, you better be darn sure that you're right. I mean that you feel very strongly about what you're putting out before you put it out because this guy behind you is going to have to live with what you've given him. (Interview)
>
> : : :
>
> Certain individuals just feel like they can do whatever they want, regardless. They don't feel like there's any need for continuity. They don't feel like if the chances decrease that they should still hang on to it in case the next model run goes back to a stronger system. They see we've got a 50 percent chance of rain, the new model run comes in, "Oh, it doesn't look like it's going to happen," and they pull it. Next model run comes in, it's back there, we put it back on. Next model run comes in, it's not there, they pull it out. Once somebody decides to put it in, ride it out until you know it's not going to happen. (Interview)

This is a problem in practice when forecasters struggle to decide whether to change a colleague's forecast:

> Randy is trying to decide what to do about Evan's midnight forecast predicting snow, which he considers unlikely. He says to Byron, "I'll have to shift where he has his significant accumulation. . . . I hate to pull that out." Byron says, "Yeah, the yo-yo effect." Sid comments that he doesn't like that Evan was vague, but notes that there might be "significant accumulations"; he adds, "I don't believe in it. It's the worst way. Why get pinned down?" He argues that until one is certain, a major change—one that the public will notice—should not be made. Randy is called by a neighboring office and says, "he inherited a watch. He's not too sure about that. He's going to keep it for the bottom two tiers of counties. He said that if he didn't have the watch out, he wouldn't put it out. He said he wished he didn't have that watch." George asks Randy, "Do we still have the same forecast or are we changing our minds slowly." Randy

responds, "We still have the same forecast, but we're making some changes. [Evan] said significant accumulation [in the Chicago metro area], I'll just switch it to significant accumulation possible. . . . My feeling is to not change the forecast too much. Not box it in. Just let the forecast ride for now. I backed off a little on the amount." (Field notes)

Randy searches for a workable compromise between the forecast of his colleague and his beliefs, and so keeps the "same" forecast, with changes. Meteorologists do not freely create the forecast that they might do *in the absence of* previous forecasts, but they participate in a system of shared, dynamic interpretations in which group relations and the public reality of the distributed forecast provide constraints. The internal negotiation, expressed openly in this case, is striking. It is not that the previous forecaster was *wrong*; he might have been correct given the information available. The question is how to incorporate subsequent information. In nonroutine circumstances such as the prediction of significant amounts of snow, the decision has consequences because external audiences respond to the forecast. The City of Chicago may put additional crews on shift, O'Hare airport may delay flights, and school systems may cancel classes.

Some choices are socially embedded, as when forecasters believe that colleagues, committed to a particular forecast, may flip-flop if they make a change. Under such circumstances, they consider their words carefully:

> Stan is deciding whether to eliminate a "chance of showers" from his forecast, feeling that good dynamics for rain do not exist, but eventually he decides not to, commenting, "I took it out two days ago, but Marty put it back in. I'm not taking it out again." (Field notes)

: : :

> Sean considers adding a gale warning for Lake Michigan, but realizes that Sid, his replacement, will likely eliminate it. He says, "I'm not going to put in gales, because Sid will just take them out tonight. I'll just put them [winds] at 30 miles per hour, and then tomorrow at nine, we can put in gales, so we don't play flip-flop games." Had Sean another replacement his forecast might be different. (Field notes)

These forecasters are not persuaded by their colleagues, but they recognize organizational virtues and power dynamics—including consistency, status, and the commitment of colleagues to their claims—as more important than precision in a system that depends on a collective presentation of the future. Forecasters perceived as too willing to change another's forecast may be deferred to, even while being criticized privately.

Meteorologists distinguish between the extended forecast (from day three to day ten) and the short-term forecast in how revisions are treated. Short-term forecasts are more likely to be changed than extended forecasts, in part because forecasters believe that they have the *right to know* what might happen in the next 48 hours. After that, prediction becomes, in a phrase often heard, "a crap shoot." Given this, if the extended forecast is to be changed, that change will occur on the day shift. The night shift typically continues the forecast, unless there is a compelling meteorological development. As I noted in chapter 3, meteorologists often rely on forecast models for their long-term forecasts, concluding that the predictions are doubtful anyhow. They select or blend the models, but often they do not assess the temperatures or probabilities of rain five days later, separate from the model. This contrasts dramatically with the effort of producing forecasts for the next two days. As one meteorologist explained about his casual approach to the model forecasts: "It's too much trouble. I don't like to change the extended forecast. It looks ridiculous. [Rain four days later] might come in early. If it does, we can change it later. . . . Unless you see a major difference [in the extended forecast] you should keep it consistent" (Field notes). Of course, each day shift inserts a new forecast for the final day as days march onward, but, given that forecasters are doubtful of their ability to predict a week ahead, they often defers to the model guidance. It is only when that third day forecast passes the gate to become the second day that the full force of meteorological attention is evident.

Leaving or Changing. The extended forecast is a special case of the more general problem of the extent to which one should change the inherited forecast. Ultimately the question is whether the goal of operational meteorology is to distribute the best forecast that one can or to have the office distribute the best set of forecasts that it can. Does the desire for a smooth and consistent set of forecasts trump the idiosyncratic insight of a particular forecaster? If forecasters own their words, and the ideas that are attached to these words, altering a forecast can involve social strain and power dynamics. While some forecasters contend that once they leave the office they have no interest in how their work is massaged, others are less open to changes. For some the temporal boundaries of worklife permit a distancing of one's self from one's forecast, but others are committed to their work products in the office or at home.

While discord is rarely explicit, on one occasion sharp conflict emerged on what seemed to me to be an exceedingly trivial issue. A meteorologist had written that the forecast for the next day was going to be "mostly sunny." His replacement, looking at the satellite images, comments, in his presence, that "There's not a cloud in the sky within 300 miles," implying that "mostly" was in error. The insulted forecaster, drawing on a history of resentment, tells me that his critic is "always talking down to me" and that if this man were standing within reach, he would have "popped him." He exited and calmed down, but he took the remark as hurtful. Of course, not every forecaster or every pair would have seen the comment as offensive, but this underlines the sensitivity of competency contests. Status hierarchies are continuously built and modified.

Forecasters debate the proper balance of deference and insight:

> [When should a previous forecast be changed?] How does it impact the customer? If the customer would have to change what they wear when they go outside or take an umbrella with them—any change like that. . . . If it's minor, I don't think the general public can tell a five degree difference in temperature. . . . Everybody takes ownership in their product. When they leave at the end of a shift, they feel that they've done the best that they can do. If somebody walks in, especially if they just walk in, they've only been there an hour or two and they say, "This isn't right," and they change it. They say, "I've just spent eight hours sweating over this forecast, and you walk in and in an hour you change it." I think people should have some consideration. . . . They need to work these things out to keep out the conflict, because it's just another thing that's just going to add to the stress. (Interview)
>
> : : :
>
> Unless you had substantial new information or a particular event that did happen between the forecasts, you would basically run with what the other individual put out, only out of respect for his professionalism and his personal integrity. You just had that understanding between each coworker. Because if you make it a habit of changing something as soon as an individual walks

> in, all you're saying is that you don't believe this person. You don't respect him. You don't think his credibility is there. It's just professional courtesy. (Interview)

In practice, most forecasters defer to the work of others, particularly if the other might become upset or has more status. So, Sean, a younger forecaster, makes a point of telling Bert, a senior forecaster, "I was going to pump up your winds, and then I said, 'I'll let it ride'" (Field notes). The politics of respect overrides one's own expert judgment. Like many knowledge workers, their hidden insecurities may infect judgments of personal competence; they often feel at risk of embarrassment or rejection.

Sometimes the desire to change forecasts triumphs. After all, they are paid to be right. Forecasters must balance their commitment to accuracy with the desire for smooth social relations. Changing a forecast is easier if one does not assert a claim of ownership:

> Almost every time, if it's something like they used "fair" [a term he doesn't like], and whenever I do a normal update, I'll change it to whatever word I feel is better. Cosmetic things like that I don't have a problem changing at all. I won't make corrections or amendments immediately upon taking shift just to make those changes, but if I have to make changes for something else, then I'll feel free to change that wording. . . . As soon as I leave the office it's no longer my forecast. They can do whatever they want with that. . . . I've been paid by the National Weather Service to put out the best forecast that I know how to, and inherent in that process is . . . the process of selecting the text that I feel conveys what I feel the forecast is going to be with the most clarity and the most competency to my constituents. (Interview)

: : :

> As soon as I come on shift, and the other guy says, "You got it? I'm leaving." It's my forecast. At that point if it were me leaving, I don't feel bad about whatever he does to the forecast, and I guess I expect the same from him. It's my forecast at that point. Maybe I didn't make it, but I'm responsible for it. If it's not working out, I don't say, "Well, he gave me that forecast." I've got to make it right. . . . I think the bigger issue is what did you know that I didn't know? (Interview)

Even for these forecasters there are limits. One claims that he wouldn't change a forecast immediately for minor issues and the other insists, particularly in the transition to evening shift when new information is scarce, that the forecaster should defer unless confident of the change.

On occasion the departing forecaster may explicitly sanction a change. For example, when Byron leaves after a complex forecast, he encourages an update, "That's why we have evening shifts [to correct the day forecast]. I hope you have a big shovel to shovel out this mess" (Field notes). Another forecaster says to the evening shift, "I'm sure there will be pressure to issue something before the ten o'clock news. If there is consensus, go ahead and let it rip." These forecasters announce that they welcome change, suggesting that in the absence of these statements changes would have been more problematic. Even if the norms of science assert communalism, this does not erase status concerns of individuals and groups. When changes are made, their significance is downplayed as when Don comments that he didn't change the forecast, but that "I tweaked the temperatures some" (Field notes). In this collaborative world, a tweak is less threatening than a change.

An office whose goal is the production of knowledge must establish coordination rules. As individuals arrive and depart, proprietors of these knowledge pools shift, and their preferences and understandings may differ. As noted, the question is whether the goal of operational meteorology is for each individual to distribute the best forecast, or, alternatively, is the goal of the office staff—a *knowledge collective*—to distribute the best forecast that they can? This choice may result in different forecasts. Given that the goal is simultaneously collective and individual, based on the belief that the two *should* cohere, the stress can be real. When an office is functioning properly, both negotiation and deference must be publicly displayed.

One Weather

I have treated the production of a forecast as if each local office operated independently. Imagine an archipelago of 122 islands of weather with no overlap.[5] Such, of course, is misleading, even if, in practice, an office operates without considering what surrounding offices are predicting when the weather is calm. The negotiated order of collective knowledge, however, links offices. Offices are connected as a network of small groups.

Under the current organizational structure American weather is a function of decisions made by autonomous offices, creating problems for those who live on the boundaries of these forecast areas.[6] One can only imagine the rainbow had the Department of Homeland Security 122 independent offices, each offering their color-coded assessment of threat.

While office autonomy is considerable, there are limits. When the forecast is complex or if severe weather is threatening, a forecaster may desire to know what surrounding offices have concluded in order to adjust the local forecast (although almost never for the extended forecast). Because the movement of weather systems is typically from the west and south, forecasters are most likely to contact offices in those directions. The Chicago office will call the offices in St. Louis, Lincoln (Illinois), and Davenport, and less often offices in Grand Rapids, Indianapolis, or North Webster (Indiana). Those offices call Chicago. Milwaukee and Chicago call each other. Through this contact, veteran forecasters become friendly acquaintances, even if they joke about (or sometimes scorn) the contrasting office cultures. Some offices, because of their traditions, networks (who transfers where), or similar meteorological concerns, have good relations. Chicago and Milwaukee consult frequently; both are older offices and several forecasters have worked in both places. Likewise the two Iowa offices, Davenport and Des Moines, work closely together, at least in part because the Des Moines and Cedar Rapids television markets cover both offices' responsibilities.

When there is "no weather," coordination[7] calls are not needed, even if the adjoining offices disagree. Sometimes temperature forecasts or likelihood of precipitation differ widely and can become troubling if, as in Iowa, media markets overlap. Forecasters may rely upon different models and often do not merge their forecasts. The exceptions are those stormy moments:

> Today Marty coordinates with Lincoln, Davenport, and Milwaukee. He explains to the forecaster in Davenport, "I'm going with an advisory. [The Milwaukee forecaster] talked me into it. I don't see any big snows coming out of this." After the call Marty explains, "I was thinking of a watch, but everyone else is going with an advisory [short of a watch]. Everyone else is, so I'm going along." (Field notes)
>
> : : :
>
> Hamilton receives calls from several local offices, primarily on whether there should be a wind warning or a wind advisory. He is thinking of a wind warning within his area of responsibility,

> even if the forecasts would not flow smoothly into each other. Finally, after talking to several offices he tells me, "For coordination, for blending the forecast, I'll go with [the advisory]. I'll see if I can convince myself." When another office calls, Hamilton tells his partner, "He's happy with the advisory, . . . but he said, 'I have some adjusting to do' " [to make his inherited forecast consistent with the forecasts of his neighbors]. (Field notes)

These forecasters see coordination in the face of storms as sound professional practice, indicating their competency in producing a collective (and plausible) forecast that blends both individual perspectives and office autonomy.

For more routine coordination, the area forecast discussion may be relied on, and once was relied on more frequently. The AFD is the account that forecasters provide, prior to their forecasts being issued, that announces their meteorological rationale. But these messages are now sent out so close to the distribution of the forecast that there is little opportunity to adjust a forecast as a result. Some forecasters read these short texts (typically 100–500 words) from other offices, but, unless there is significant weather, they do not respond. The discussions are available to media outlets and are now placed on NWS public websites so weather weenies can see what forecasters are thinking.

The area forecast discussion generates tension because of its multiple and distinct audiences. Forecasters are supposed to distribute their discussions prior to their forecast to allow for coordination with other offices, but they are simultaneously distributed to the media to permit these translators to plan their patter. In the first case the communication is subcultural shoptalk, in the other it is didactic. A forecaster in Belvedere takes the coordination aspect of the AFD so seriously that he attempts to issue his by noon while still deciding what to forecast, so neighboring offices can read his thoughts about the trends in the current weather system. He is an exception; most forecasters typically select their forecasts and then may examine these discussions to check that they are not misguided. One Chicago meteorologist produces his forecast in reverse. He composes his forecast, and only then writes an explanation for his decision. His AFD is typically distributed after his forecast, preventing it from having any role in coordination among offices and used only by broadcasters. The fact that he hasn't been penalized for his practice suggests the marginality of the AFD as occupational shoptalk. To encourage coordination, the NWS has recently established a computer chatroom to permit forecasters to share ideas. One critic of the AFD explains:

> I hate them . . . I'd like to get rid of them. They used to be a coordinating method between the weather service offices, and they were good. You could tell other forecasters in technical terms what you were thinking, and you could coordinate. That's what they were for. They were internal. But some bureaucrat decided that they would have to go out in public, so the TV meteorologists can read what they are. It's not my job to teach them. So how I write them now, I write them in English. I focus on the big features only. (Interview)

The prominence of the end user influences the forecaster's attitude.[8] Is this collegial communication or is it aimed across the boundary of scientific brotherhood to an uninformed broadcaster (or worried public)? As I discuss below, this expectation affects writing style.

Negotiating inside the Box

On most days from spring to fall, forecasters at a few local offices will receive a call from the lead forecaster on duty in Norman, Oklahoma, priming them for the possibility of severe weather. Before the Storm Prediction Center issues a *watch box*—a parallelogram informing the media and, thus, the public of potential severe thunderstorms or tornadoes in a particular geographical region—they inform their colleagues. These boxes are now widely known for being routinely displayed on the Weather Channel. The lead forecaster at the SPC is required to contact the local offices inside the box. These coordination calls suggest, however, that the decision should not be hierarchical, but collaborative, negotiated between the storm center and the local office.

The two individuals on the phone—and their organizations—have different concerns, and, as a result, a possibility of discord exists, although usually the disagreement is polite. The SPC has the responsibility for the entire nation and makes decisions without regard to local impact. Their goal is to determine which weather systems might produce major storm outbreaks. The goal is to draw watch boxes that specify the location of these potential tornadoes and severe thunderstorms with plenty of time to allow local offices, media outlets, and citizens to prepare. But each watch box has organizational consequences. In many locations, a watch activates local emergency units. In some counties, as long as a watch exists, emergency offices must be staffed. Local forecast offices also feel a watch's effects. If a watch is issued, forecasters may be required to stay late, come in early, or be brought in specially, with budgetary consequences for overtime and in disruption of family life. When a local office is short staffed, they may convince themselves that what they desire for organizational needs is what they see. If the local office agrees that severe weather is likely, conflict is avoided, but this is not always the case. Add to this the belief that offices have different "philosophies," exemplifying local idiocultures. One SPC forecaster listed offices that "will warn on every blip," noting that "some of the offices in large metro areas would be more passive" (Field notes).

Watches often are drawn to the edge of a local forecast office's area, state line, or media market, although weather ignores such bureaucratic niceties: "We can bring it down to the Arkansas border" or "We'll keep Indiana out of it" (Field notes). In this sense boxes are political entities; boxes that follow office or state lines are "clean boxes." Forecasters describe these as "geopolitical decisions." Watches are more than meteorological choices; they affect people inside and outside of the weather service. One forecaster was explicit that the effects of his boxes were on his mind: "Boxology is a big issue, and it's something that you have got to be thinking about and I ran into a little of that this weekend. I tried to minimize what I did in boxes where it affects the WFOs [local forecast offices]. We can say that the meteorology doesn't know borders. That's true, but it can certainly make it easier for the forecast office whenever you can" (Interview).

These consequences are also evident in the practice of drawing clean boxes:

> Guy draws a watch box for Missouri and Illinois. He comments about his construction, "I'll try to make it as clean as I can. I'll include Lincoln's CWA and the rest of St. Louis's CWA, but I won't go into Indiana [and thus won't involve another office]." Building on that box later, he tells the Paducah office that he will include all the counties up to the Tennessee border, the end of their forecast responsibility, adding "It might be cleaner that way." (Field notes)
>
> : : :
>
> Terry issues a severe thunderstorm box for southern Missouri and southern Illinois. He concludes that the eastern part should be a tornado watch but wants to avoid including the St. Louis

> forecast area. He muses, "I thought of taking the southeast corner as a tornado watch. I don't know the best way to handle it. . . . I kind of feel bad. [One of the St. Louis forecasters] asked about a tornado watch, and I said, 'We'll see how it develops,' and now I'm making more work for him. . . . I feel bad because St. Louis is going to be busy, dumping this [second watch] on them. St. Louis is going to think I'm Mr. Upgrade." [He had upgraded a thunderstorm watch to a tornado watch the last time he was on duty.] (Field notes)

The SPC sees coordination as a means to manage their colleagues, but they typify offices in their local orientation, rather than by the preferences of individual forecasters. I was told that one of the senior forecasters knew the preferences of each *office* and tried to accommodate them: Do they prefer early watches or would they rather wait? Senior forecasters at the SPC had advantages by virtue of being able to refer to their experience, but every forecaster recognized that negotiating with local offices could be contentious. While SPC forecasters have the authority to issue watch boxes over local objection, these forecasters could then eliminate counties from the box or, after the fact, if the SPC forecaster was wrong, complain.

During my two weeks at the SPC I listened as these negotiations took place, sometimes between an SPC forecaster and a local office and sometimes on a conference call with several offices. Deference was typical, as the assumption was that they were colleagues searching for truth and preserving public safety: "Would it be OK to put in all your counties. Do you have any problem with that?" The goal is for the conference call to demonstrate teamwork through discourse. The emphasis on collaboration attempts to invest local offices in the decision, but the strategy requires effort, potentially leading to inefficient decision-making or creating conflict.[9]

The conflict need not be between only the SPC and a local office but among local offices. In some conference calls local offices disagree. In one call the Springfield, Missouri, office wanted a thunderstorm box and the Paducah, Kentucky, office preferred a tornado box. While two small boxes could have been issued, this is seen as confusing, hard to describe, unaesthetic, and unprofessional. The Springfield office finally acquiesced to a tornado box. At times media concerns influence a local office's preferences. The Fort Worth office asked to have the boundary of their box stop outside the counties of their Metroplex because of perceived media pressure.

Negotiation can involve issues of timing as well. A box justifies overtime, and so a box issued well before the likelihood of severe weather may be organizationally desirable. Forecasters at SPC believe that early notice demonstrates respect, and staff must balance their certainty (the earlier, the less certainty) with a desire for timely notification.

A region that has recently been hit by major severe weather is, for a time, willing to acquiesce. Less than a week after the deadly La Plata tornado in suburban Washington, severe weather again threatened. While two other offices resisted a tornado box ("neither one of them wanted to go red," the color of a tornado box), the local Sterling office accepted. As one forecaster said, "At this point, they can't say no" (Field notes).

One strategy for building support is through *priming*, notably by means of the severe weather outlooks that the SPC distributes for the following three days. Placing an area in high or moderate risk for the day or the day ahead sensitizes offices to the possibility of severe weather. These offices are now cognitively and organizationally ready for severe weather. The same system that might have been considered annoying is now labeled potentially destructive, and the local office becomes more agreeable. When an office is not primed, they may resist until the SPC pushes the point. One SPC forecaster relates, "Nashville is shocked. They were not expecting severe [weather] tonight." His partner responds, "Well, at least we got his attention. He couldn't believe that they were going to get severe weather. He was flabbergasted. Will he have to call someone in? I hope something happens." The Nashville forecaster accepted the watch, and it turned out that the area received a strong tornado near dawn (Field notes).

Sometimes offices object to proposed watches, requiring delicate negotiation to maintain organizational comity. As noted, the Chicago office is meteorologically conservative, desiring to avoid watches and warnings unless absolutely necessary. Chicago forecasters believe that the SPC becomes overly excited about small risks. (SPC forecasters see *themselves* as conservative, only forecasting major outbreaks, ignoring isolated storm systems. As one explains, "we have resisted as hard as we can this *watch inflation*" in the face of pressure from headquarters and from media outlets, such as the Weather Channel, that use watches and warnings to capture viewers.) When calling Chicago, SPC forecasters know that they may need persuasion and feel relieved if a sympathetic colleague is on duty. I listened to a call to Chicago in which the local forecaster didn't want a watch, then agreed to it as long as it included the entire Chicago area, to which the SPC forecaster acquiesced. The SPC forecaster explained, "If you feel strongly you don't really need it,

I'm not going to force it." Apparently the Chicago forecaster attempted to persuade the SPC to include Wisconsin, leaving out Illinois. When he gets off the phone, his partner asks whether the Chicago forecaster was really critical and is told that "he wasn't belligerent," a reference to a time past. A few days later another SPC forecaster considers asking if they could include a few southern counties in a watch, feeling that those counties might be at risk. He comments to his partner, "That will be a hard sell." Fortunately, an accommodating forecaster was on duty, and the watch was accepted. Chicago was not the only office that resisted watches and was not the only office so typified. But the fact that the Chicago office is seen as being resistant suggests that workplace negotiations are not only with individuals but with group cultures.

Ticky-Tacky Boxes. As I discuss below, local forecasters are wordsmiths. In contrast, at the Storm Prediction Center in Norman, Oklahoma, words matter far less than their beloved boxes. Although one might imagine that the forecasters simply draw (on a computer screen) as they wish, focusing on heightened meteorological risks, this communication is tied to social concerns and a sense of occupational aesthetics.

As I have noted, watches were originally designed for radio, established so that broadcasts could report that a tornado watch exists *x* miles either side of a line from *y* to *z*. This announcement created the visual image of a parallelogram or box, and this remains the standard means of communicating severe weather threats. Even as more and more people are informed through visual media—television and computers—boxes have remained. (The weather service is in the process of changing them to polygons to make them more flexible). The former technology has determined how public safety is communicated, even after technological change.[10]

While boxes could be any size and shape, meteorologists embrace an aesthetics of "boxology." Forecasters do not like to draw boxes that are too large or too small (a "postage stamp"), or too squished (long and thin). By putting the storms in a box, forecasters hope to allow for the storms to move through that space, minimizing the number of new boxes. In addition, boxes do not stand alone. One forecaster explained, "You want to put your boxes where you can build on them. You're constrained by the rectangle quality. You can't put in a triangle watch. . . . This is what the truth is [he draws me a small shape on the map], but it would make it hard to build on to" (Field notes). In other words, one must not only be concerned by what is happening at the moment, but what might happen in the future. Discussion about the proper shape of boxes is common:

> Lawrence is considering how to draw a box for storms through North Carolina and Virginia (eventually they draw two boxes for the region). He comments, "I wish we could do an L shape." Terry jokes, "I wish we could do an inverse Oklahoma [a box in the shape of Oklahoma]. If I could do a polygon, I could take care of it."[11] Nate says, "You know what I'd do. I'd make it last for an hour and a half [a short period for a watch], and then I'd give it to Guy. 'Guy, it's your mess now.'" Terry comments, "I hate to overlay one watch on another. Too bad we didn't shift change at three o'clock [so Guy would be responsible]." (Field notes)

: : :

> Numerous boxes have been issued today. Stu sighs, "What a mess, boxology-wise." He tells Guy: "It may be a waste of a watch, but could you extend this box farther to the east." Guy responds, "Maybe I should just sit on it [waiting until the storms move and form a better shape], but I sort of hate sitting on this storm. It looks pretty tough." Later Guy is talking to the Memphis office: "It's not real clear what we need to do to handle these storms. If I issued a little postage stamp for that, it would handle those storms, but it would not really be appropriate [being too small]. I was thinking of extending the watch for your Arkansas counties farther east." Later Guy talks about issuing a small box for Kentucky but says to the local office, "There is so little room there [between boxes] that I can't issue a watch. I would be comfortable if you guys handle it warning-wise, and let it expire, and at the congressional hearing [if disaster occurs] that's what I would tell them." (Field notes)

On rare occasions postage stamp watches are acceptable, but they demand a strong justification. Borderline severe weather is insufficient.

While boxes are linked to meteorological events and are not merely artistic or political expressions, how they become formulated is tied to occupational norms. As forecasters are socialized, they are taught to communicate in ways that exhibit competent professional practice. Squashed, sheared, flattened, narrow, or overlaid boxes, while sometimes necessary, indicate that the forecaster has not thought ahead sufficiently. These reveal that the forecaster is only now "catching up" on predictions that should have been made earlier. They are spoken of as "ugly" or "messy." It is because of the subcultural components of watches that forecasters speak of boxology, an art and a science as compelling as meteorology itself. The prediction of what will happen on the ground should fit standards of efficient and competent practice. Seemingly mundane, these communicative technologies are ultimately evaluated through internal standards of which outsiders may be only dimly aware.

The Write Stuff

Just as operational meteorology is a collective enterprise, it is also a literary exercise.[12] In this, as many scholars argue, notably anthropologist Bruno Latour and economist Donald McCloskey, it stands with other disciplines—a world of persuasion based on language.[13] The product of the scientist is words. Said a forecaster, "We're word craftsmen." Perhaps one should not push this too hard as forecasts rely upon a limited number of words, a paucity of nouns, adjectives, and adverbs, and near absence of verbs. These texts do not quite reflect Latour's assertion that science is "thrilling," a "real opera" with heroes and villains;[14] however, they do contain drama, even if the style is spare and bony.[15] The writing of operational meteorologists is closer to a *literary technology*, as discussed by Steven Shapin,[16] an instrumental resource shaped for persuasion. Ultimately the forecast is nothing if it doesn't provoke its audience to act. Weather forecasting, whether directly from government forecasters or massaged by the media, is one of the most powerful and insistent forms of persuasive communication in all the sciences.[17] Consider the forecast for Chicago (technically for McHenry, Lake, Kane, DuPage, Cook, Kendall, and Will counties in Illinois and Lake and Porter counties in Indiana) for the week beginning the evening of Friday, January 11, 2002:[18]

> TONIGHT . . . INCREASING CLOUDINESS TOWARD MORNING. LOW AROUND 30. SOUTHWEST WINDS 10 TO 15 MPH.
> SATURDAY . . . CLOUDY. A 40 PERCENT CHANCE OF LIGHT SNOW OR FLURRIES. HIGH AROUND 40. SOUTHWEST WINDS 10 TO 15 MPH BECOMING NORTHWEST 15 TO 20 MPH.
> SATURDAY NIGHT . . . CLEARING. LOW IN THE MIDDLE 20S. NORTHWEST WINDS 10 TO 20 MPH.
> SUNDAY . . . MOSTLY SUNNY. HIGH 35 TO 40.
> SUNDAY NIGHT . . . BECOMING CLOUDY EARLY WITH LIGHT SNOW LIKELY. LOW IN THE MIDDLE 20S. CHANCE OF SNOW 60 PERCENT.

MONDAY . . . LIGHT SNOW LIKELY. HIGH 30 TO 35. CHANCE OF SNOW 60 PERCENT.

MONDAY NIGHT . . . CLOUDY WITH A CHANCE OF SNOW EARLY. LOW IN THE LOWER OR MIDDLE 20S.

TUESDAY . . . PARTLY CLOUDY. HIGH IN THE MIDDLE 30S.

WEDNESDAY . . . CLOUDY WITH A CHANCE OF SNOW. LOW AROUND 20. HIGH AROUND 30.

THURSDAY . . . A CHANCE OF SNOW SHOWERS. LOW AROUND 20. HIGH IN THE UPPER 20S.

FRIDAY . . . A CHANCE OF SNOW SHOWERS. LOW IN THE TEENS. HIGH IN THE MIDDLE 20S.

This is a fairly typical midwinter forecast for the Chicago area. No severe weather is forecast, no frigid temperatures. Several things are notable about the week-long forecast (actually seven and a half days). It is a signed document. I know who wrote it: "RLB," one of my informants. Meteorologists sign their forecasts. Sometimes with their last name, sometimes with their initials, and sometimes with an identifying number—one forecaster chose "86" to honor the 1960s television sitcom, "Get Smart!" and to signal his affection for the main character.[19] At the few offices that do not require signatures, some feel that their forecasts are depersonalized. Having one's name attached to one's work reinforces the belief that "you should be proud of your product" (Field notes), but also that the organization considers forecasters accountable for their predictions. While the public does not know these names, coworkers in other offices and media personalities may. The signature prevents the forecast from being purely a collective product, tying it to the self of the scientist. To be sure, this document owes a lot to the previous forecast, to yesterday's forecast, and beyond, but even if the forecaster borrows words, they are his, even while being transferable. In operational meteorology one person cannot plagiarize the forecast of another.

The forecast uses 119 words for a week's worth of weather (not counting numbers or days of the week). There are thirty-five separate words, six nouns, and no true verbs (although increasing and clearing could be so considered). Only four—"increasing cloudiness" and "becoming" twice—have more than two syllables. This is a text that in its level of writing could teach second graders how to read. And for children the words have the benefit of demonstrating that experience can be revealed through text. Indeed, forecasters know that their words will reach the public, either through the Weather Channel or on their own website, typifying their readers as "Joe Blow" or "Joe Schmo," a (male) reader with a fifth-grade education.

Yet, there is also considerable subtlety. Not in the words themselves, nor in a sense of style that dramatically differentiates forecasters from each other, but in the ability to link words to the events to which they are to refer. Cultural knowledge is needed to unpack the text, not only in knowing the denotations of the words. How do "flurries" differ from "light snow" with which they are contrasted? What are "snow showers," and how do they differ from the rain "showers" that are their counterpoint.

The connotations of meteorological writing are seasonal. Terms such as "cold" or "warm" have different meanings in July, September, and February. One September forecast was labeled cold with temperatures in the 30s. That day might have been described as "mild" four months later. Temperatures in the 50s have different meanings in February and July, and they are referred to in dissimilar ways.

It is the assumed transparency of language that makes this rhetoric so effective. It feels *rhetoric-free*. One would have to strain very hard as a "cultural dope"[20] not to understand this text. One could do so, but only by reminding oneself how often in practice writing lacks ambiguity.[21] Each day two sets of forecasts are distributed—typically with three or four variants for different "zones" or sets of counties. During this research, the Chicago office usually issued a zone forecast for the Chicago metro area, one for the northern and western counties, one for the southern counties, and sometimes one for the southwestern counties. The number of zones is a function of the complexity of the forecast for the next two or three days. After that the extended forecast, less certain, tends to become similar for all zones. Chicago may issue eight zone forecasts in a typical day, and 121 other offices do the same. To be a meteorologist, at least at the moment of this research, was to be an author, a wordsmith who creates multiple texts.

The Meteorological Self in Words. Every occupation involves an *identity game*. How you work affects who you are. Work creates a pattern of salient self-images. The elements that are selected to represent that self are, however, a function of the occupational culture. All workers—clergymen, tailors, or muggers—must communicate with their publics and so labor is a linguistic game. In meteorology, words are a heavy—and startling—part of this process. Partly it is through words that workers differentiate themselves from machines, giving priority to human authority. These words, formulaic as they are, matter to the forecasters.

Writing a forecast is, in some sense, a form of emotion work, both in that they feel strongly about protecting their fellow citizens and in that they desire to be unconstrained in their communication choices. They are not overtly concerned about style as such, but their predictions matter, and their predictions are expressed in words. From my first days of observation I was told about the importance of the ownership of the forecast and its words. It is for this reason that meteorologists were concerned about the changes that computerized forecasting would bring, altering both what they do and who they are, modifying their autonomy to communicate with imagined audiences. As I discuss below, this new system, requiring forecasters to manipulate databases, creates the wording directly from the data manipulation without literary input from the forecaster. My study examines forecasters at a historical moment, focusing on one office, although the concern with wording was also evident in the two other local offices. Because of the timing of the study, I cannot assess the attitudes toward writing after computerized forecasting was fully accepted and after a new generation of forecasters were socialized to that system, but even after forecasters were using this new system they routinely edited the computer forecasts and talked about their wording choices. One MIC explained: "We're authors of our own products each day. So each person likes to have their personal touch in their writings, just like you write your articles. These people write their short articles each day based on their judgment, and, so, it becomes a very personal thing, the wording that our people put into it. And now, if we're going to go into a [computerized] system which tells us what the words should be and many times these words will be the same as what we had before, but now *some machine is telling us what to say*" (Interview; emphasis added). One of his forecasters makes this belief evident in striking language, considering what outsiders might find to be the constrained nature of the texts: "Being able to use words that are as descriptive as possible . . . not flashy but in an artistic fashion that either delight or make it easier for them to understand. . . . Since I've been doing it for twenty-some years, it's become an integral part of me. . . . And to take that away, say 'You can't do that anymore,' it may not have seemed important at the time, but now seeing something being taken away, your job's changing. I do kind of feel resentful" (Interview). Put another way, "People feel they are losing all their individual artistry. . . . They are worried about losing their own words," "Sometimes it becomes artwork. We take these numbers and massage them into words," or "It takes away the art. It's our product. It's part of us. We don't want a machine to tell us what to do. It's an insult" (Field notes). Workers feel that they lost turf, relinquished a valued task and yielded the imprint of their vision on their work products.

It is not only the control enforced by computer forecasts that threatens the autonomy of the forecaster as *auteur*, but so do organizational choices.[22] During summer 2001, forecasters waited with trepidation for a revision of "Chapter 11," the document published by headquarters that told forecasters how they must format and phrase texts. Despite the desire of forecasters to control their own writing, the ultimate decision was bureaucratic, even though local forecasters were permitted some leeway. Every so often headquarters revised the style and content that were permissible. This revision was not as dogmatic as feared, but alterations were mandated. The new version of Chapter 11 outlaws writing "mid to upper 40s," but they could still specify a five-degree range, such as 44–48 degrees. One forecaster explained, "In one respect it takes away our stuff, but it's nice to have a consistency. It's nice to know that the words that are used are the same." But his colleague jokes, "No one is following it anyway. . . . They just ignore it" (Field notes).[23]

The computerized forecast has a more obdurate reality. Ironically given the eventual decision of the Weather Channel to create their own wording, it was TWC's insistence that led to the pressure from headquarters to present forecasts in a standardized format.[24] With each attempt at standardization, forecasters worry about lost authority. The trade-off, at least as imagined by headquarters, is that better data and more complex models permit improved forecasts, and, eventually, higher public esteem. Any worker who has the skills to manipulate such complex data could surely imagine themselves as a competent professional.

Many forecasters are not convinced that the public notices changes, but they claim that the changes affect the public through the decay of language, as well as constituting an attack on who they are and what they do. In the words of one, "The words carry specific meaning. You spend a lot of time figuring out what you mean." A forecaster may spend up to an hour editing texts, such as polishing "outlying areas" to "outlying suburbs." Forecasters frequently discuss their selections with colleagues. This was particularly common as younger forecasters asked senior colleagues to proofread or asked whether to include "scattered showers" or a "chance of showers."[25] Even though they have a hazy view of their public, forecasters like to believe that the public reads as carefully as they write, striving to give the best advice. Local offices imagine an audience much like themselves. This is a standard problem for many occupations—whether pharmacists or software developers—who hope to communicate their insight, but can only imagine audiences with their own concerns and sophistication. This belief opens a space for the media to translate their words into forms that are more accessible and usable for other audiences.

As emphasized by their signature, meteorologists own their forecast. As one put it, "You can storm chase all you want, but until you put your name down, it doesn't count" (Field notes). Another comments, "You have to have a thick skin and a sense of humor to be a meteorologist. You put yourself on the line whenever you put words on paper" (Field notes). The importance of the signature was evident in a jocular conversation between two meteorologists:

> DON *(joking with Byron)*: I put your name on [the forecast].
> BYRON: In that case, I'll make some changes.
> DON: You think I want to take the blame for this. *(Both laugh.)*
> BYRON: I won't send it out blindly.
> BYRON *(later, to me)*: Let's hope this forecast is right. He has his name on it. (Field notes)

For an agency that stresses communication between the forecaster and the public, exhorting forecasters that "the 'words' is what we do," a change from this system proved traumatic.

Even the changes from the written forecast to the media representation of it can be significant. One forecaster mused, "You stand in a line at the grocery store, and people will say that they are predicting this. You feel like saying, 'I'm they.' You'll write a forecast and you know what you wrote. You'll write a 30 percent chance of rain toward the evening, and the DJ says, 'Rain,' and you'll say, 'That's not what I wrote'" (Field notes). Even when filtered through the media, the words are still owned. As a result, poor writing, sometimes attributed to younger forecasters at spin-up offices, takes a toll. Such writing, however defined, threatens the occupation: "To me it's very important because we're seen as being educated people in this office, at least that's the way we want ourselves to come across. When you have people writing forecasts that use incorrect grammar or leave articles out of sentences, poor sentence structure, run-on sentences . . . *it just makes my skin crawl* that people are looking at that on the Weather Channel.[26] It goes out verbatim, just the way it's typed in here. You've got these hundreds of thousands of people that are reading this, thinking, 'Geez, what kind of a [person] do they have working in that office'" (Interview). The sensitivity is such that the office displays examples of the poor writing of *other* offices. Privately, colleagues could be contemptuous about each other's writing, even in one case going to the MIC to complain about a colleague whose forecast used "becoming" too frequently, which, while grammatical, was unaesthetic, or, as it was put, "really sucks" (Field notes).

Repeatedly I heard a mantra of the need to communicate with the public, a marker of selfhood. This led to a philosophy of writing. Forecasters discuss and debate linguistic choices, for instance, how to write headlines to communicate the proper level of threat. One forecaster objected to excessive hedging, using "mostly" or including small chances of precipitation instead of presenting a firm conclusion.[27] He jokes about this linguistic preference, "We always hedge. Partly cloudy. Chance of rain. It's never sunny. It's always mostly sunny. We don't want to be wrong. We can't be 100 percent right. It's never black. It's always mostly black" (Field notes).[28] Significantly a colleague held the opposite philosophy, noting, "If you get specific, the odds are you're going to be wrong. You make a choice and you're going to miss it. You don't want to get tied down." No single aesthetic philosophy is embraced, and as each forecaster has preferences, the philosophy can yo-yo, along with the predictions, unless one forecaster shows deference to the desires of another. These preferences affect the forecast words, which in turn influence citizens in their daily activities, even if their audience does not appreciate the technical meanings of "partly cloudy" or the implications of a "30 percent chance of showers." The media, filtering these claims, have their own philosophy of language. Different institutions have their preferences, a function of aesthetics, technology, or the imagined needs of their audience.

Fair Exchange. The politics of rhetoric may result in discussions—battles—over particular words and phrases. With a spare writing style, the individual word or turn of phrase is crucial in ways that it isn't in this bulky volume. Forecasters have different preferences in their pressured poetics. As one explained: "[I try] to communicate to the public exactly what I would expect to see and feel when I walk out the door. I like using subjective terms in a forecast too. For most people relate to those a lot more than just numbers. For instance, if it's cold outside and kind of a cloudy day and the wind is really blowing hard, well, I like to use the term 'blustery.' Blustery cold to a lot of people means a hat and scarf today, a heavier jacket to wear. And then, of course, 'sultry,' another good term. They communicate a picture of things. But I feel that a lot of people relate to them, as opposed to just saying today's sunny and high 90 to 95" (Interview). Not all meteorologists agree with his preferences, but everyone acknowledges that each forecaster has preferences. A debate occurred about whether to use "wind" or "winds"—as in "strong wind/s." Some felt that wind was a proper collective noun, and others felt that winds is a Chicago regionalism and complained that when "wind" is read by the automatized voice on NOAA weather radio it is pronounced "wined," an error that must then be corrected. The algorithm by which the machine determined the proper pronunciation of this heteronym made the wrong choice. The voice recognition software became another audience that forecasters had to consider.[29]

Some preferences seemed eccentric, although meaningful to the writer and often linked to his or her training. One forecaster happily used "flurries" (of snow) but objected vehemently to "sprinkles" (of rain). He says that he was told as an intern, "If it's going to rain, it's going to rain. If it sprinkles, it still ruins your picnic" (Field notes).

The most dramatic battle of words, in the Chicago office at least, was an ongoing debate over the use of "fair." Could such a widely used word be standardized, and would such usage depend on public understandings or technical ones. Whose language was it? Whose definitions triumphed? Such a debate is widely felt in many knowledge disciplines that hope to communicate with nonspecialists. Whether we discuss "status," "norm," or "fair," we must select the limits and boundaries of definitions as used in practice. Could "fair" become a technical term, or would its roots always remain in public space. If the former, would its use in forecasting create confusion among a public unwilling to cede this sturdy term to specialists.

A few forecasters embraced the term "fair" and its metaphoric quality, feeling that it should be used for a nice, mild, pleasant day or night[30] with only a few wispy, cirrus clouds. They felt that the term had a rich intuitive meaning. Others objected, arguing that fair lacked a clear or standardized meteorological value and was ambiguous. The debate centered on whether this emotive word had a stable symbolic meaning and whether such language was transparent and accessible to members of the public. Critics preferred the equally metaphorical "clear." There were rumors that "fair" would be banished by headquarters who, it was said, wanted standardized language, but, to everyone's surprise, "fair" survived. The formal Weather Service definition of fair became "few or no clouds below 12,000 feet with no significant weather and/or obstructions to visibility," creating a bureaucratic standard from a metaphor, reminding us that even the smallest matters can become matters of bureaucratic concern. This debate became enmeshed in the office's joking culture, and was learned of in other offices, such as Flowerland, where forecasters describe "fair" as "Chicago's favorite." Although there was much joking, real disagreement existed about this four-letter word:[31]

> Sid comments to Stan, fair-weather friends, "You know these fairs. The midnight crew kicked them right out, but I put most of them back." (Field notes)[32]
>
> : : :
>
> Stan tells Sid about his forecast, "It looks like a bunch of 'fairs' coming up." Sid answers, "I know, use it or lose it, buddy." Later Sid joshes, "I put in a fair and it was changed to mostly clear. Some people are not team players." (Field notes)
>
> : : :
>
> VIC *(joking about revising forecasts after the new computerized software is installed)*: Fair is not allowed.
> SEAN: It's strongly discouraged.
> VIC: Which means for this office you have to put it in.
> DON: I'll put it in.
> VIC: You know, doctor, you're an infidel.
> SEAN: You can't say poor, so you shouldn't say fair.
> DON: Just for that, I'll put in fair.
> VIC *(joking)*: That's OK, I can put out an update.
> SEAN: Increasing fairness . . . variable fairness . . . chance of fairness. (Field notes)

It is symptomatic that the younger forecasters, more comfortable with standardization, like Sean and Vic, generally oppose fair, while the older men, who value their autonomy, like Don, Stan, and Sid, like fair and other emotive writing. The proponents of fair believe that the term has an intuitive folk meaning:

> Maybe it's from being of the old school, but fair to me has a distinct meaning. Fair means that you have no extremes of temperature or wind, and the sky condition for the most part is clear. . . . Fair had definite meaning. (Interview)

Critics disagree with the claim of definite meaning, and see the terms as seeding confusion: "I'm not a proponent of fair. What's fair? . . . If you put four people in a room, and asked them what fair is, you'd get six different definitions. That's the problem with words like that"

(Interview).[33] The dispute revolves on reader response. It is not that anyone loves the word "fair," the question is—as it is with all writing—whether it communicates. Because meteorologists use so few words and aspire to use each precisely, the debate is intense, if good-humored.

Opening One's Mind. Forecasters distribute many products, both routine and pressured, including severe weather warnings, produced in a white heat, using well-established formulae. However, if one counts the number of words written, and perhaps even the time spent, the area forecast discussion constitutes the bulk of meteorological writing. It is not as important for a sense of self as the forecast, but it is narrative. In the AFD the forecaster justifies the prediction in several hundred words—the number varies as a function of the author and of the weather conditions. As I noted in discussing coordination, their audience can either be fellow forecasters or broadcast journalists (or private forecasters) and now, thanks to the Internet, the public.

These distinctive audiences create problems for the form of writing. Not too many years ago the AFD was distributed within the agency, an internal product designed for coordination, a form of shoptalk.[34] This permitted forecasters to write in clipped, abbreviated, acronym-laden language, as some still do:

> LATEST AFTN VISBL SATLT PICS SHOW SNOW COVER REMAINING OVER SERN PORTIONS OF FA SO WILL GO A COUPLE OF DEGREES COLDER THERE FOR OVERNIGHT LOWS AS DECOUPLING LIKELY EVEN WITH INCRSG SLY FLOW LATER TONIGHT. CS DECK MOVING EWD FROM IA IS FCST TO CONT TO INCRS IN COVG SO WILL MENTION ICRSG CLOUDINESS FOR TONIGHT.

Translation: The latest afternoon visible satellite pictures show snow cover remaining over the southern portions of forecast area, so I will forecast a couple of degrees colder there for overnight lows, as decoupling is likely even with an increasingly southerly flow later tonight. A deck of cirrostratus clouds moving eastward from Iowa is forecast to continue to increase in coverage, so I will mention increasing cloudiness for tonight.

The extract above is a portion of a larger text, incorporating meteorological knowledge such as "cirrostratus deck" or "decoupling," acronyms, such as CWA (County Warning Area), and various abbreviations and shortenings. In recent years the Weather Service has attempted to make these documents more user-friendly, discouraging abbreviations, defining users more broadly than some forecasters like. As one forecaster put it, what had once been a *discussion* is now a *synopsis*. Risque in-jokes, such as the acronym for "scattered cumulus clouds"—FEWCU—are largely a thing of the past. Now AFDs, aimed at the media and public, while still requiring a measure of meteorological knowledge, are less likely to require translation:

> SATELLITE IMAGERY SHOWS CLOUD COVER OVER THE UPPER MIDWEST CONTINUING ITS EASTWARD MOVEMENT AS THE UPPER TROUGH OVER THE AREA PROGRESSES TO THE LOWER LAKES. EXPECT SKIES TO BECOME SUNNY BY MID MORNING AS WEAK RIDGING SLIDES OVER THE FORECAST AREA TODAY AHEAD OF THE NEXT SYSTEM TO COME TO THE AREA.

This has less of a feel of shoptalk than the previous extract, opening up the profession to the public, altering the boundary between scientist and layman. Imagined audiences affect communicative strategies. In their newsletter the Belvedere office once decided to explain some of the more common acronyms used in their AFDs. In the Chicago office, attitudes toward the public use of the AFD are more likely to be sheathed in humor, "I might as well send all this good stuff out, and let the world feast on my wisdom. The world has my thought process. Isn't that scary?" (Field notes).

Even more than the public, the media has pressed for the AFD to be accessible. They desire a concise text aimed at a person with modest meteorological knowledge, providing them information to use on camera. This translation is less necessary for knowledgeable meteorologists in major markets, such Chicago's Tom Skilling or Washington's Bob Ryan, but useful for their colleagues in smaller markets. One forecaster told me that he writes for "weekend weather." Another imagined a comely young broadcaster as his reader: "I have Tracy Butler in mind. . . . She's the local weather girl on Channel Seven . . . she's got a degree in meteorology, but hasn't got a lot of experience in weather. She's learning it, and so I have her in mind" (Interview). Writing for novice meteorologists requires a mindset different from writing for one's experienced professional colleagues. This debate, a function of meteorology's role as a public science, is addressed in chapter 6, but it reminds us of the multiple audience problem that forecasters face.

Unlike the forecast itself, attached to the self of the meteorologist, the AFD is written for an audience, and internal fights over wording are less common. It is writing, but not from the soul. Writing has different value depending on how it is linked to the core of occupational work, akin to the differences between published articles versus internal memos in other work domains.

"If-Pissing" Life Away

Early in the 1990s the National Weather Service headquarters, besieged by private companies demanding that the service be privatized, decided that its future would be to present detailed databases of meteorological conditions that others could use.[35] An outcome of this decision was the system termed IFPS (Interactive Forecast Preparation System) through which forecasters would manipulate databases, primarily by drawing meteorological boundaries (with computer tools) on gridded maps. The computer program could then rapidly prepare a suite of products, creating wording, and these grids would be available to users—the media, private firms, or even the public.[36] The boundaries between the scientists and their publics were shifted as a function of institutional pressures, common in an age of deregulation and privatization.

To the extent that the previous forecast or the model fit the forecaster's belief, more or less work is required, as long as the forecaster accepts the wording that the computer provided for those forecast zones that the meteorologist had selected. Forecasts are not limited, however, to zones based on counties, because the forecaster creates a detailed map of the forecast region.[37] As a result, a forecast is available for each point on the map. Previously forecasts had been written for counties or groups of counties; soon one could have a forecast for a grid of one square kilometer, a personalized forecast.

Part of this change entails having the meteorologist focus on the science of the forecast, "letting the machine produce the words." The grids as drawn with gradients for temperature, precipitation, wind, and cloud cover are the meteorological product. Although this might be treated as heightening the scientific credentials of forecasters, the results were more controversial.

As I emphasized in chapter 2, the study of work must involve the examination of microcultures—in this case, the differences among the three offices. At one point after the system was well on its way to being implemented, the NWS surveyed employees as to whether they thought it would be "beneficial."[38] Overall approximately 50 percent agreed, but in Chicago the vote was 14 to 1 against. This reflects my sense as I watched the implementation process. From my observations I suspect that in Belvedere the vote would have been reversed, and in Flowerland, it might have been more equally split. Not surprisingly—and both the cause and effect of these attitudes—the system was implemented earliest and most smoothly at Belvedere. Regional headquarters was well aware that considerable opposition roiled Chicago. Indeed, the date for full implementation was pushed back several times in Chicago from March 1 until May 15, although eventually it was implemented and not found to be quite as distressing as it first appeared. Over a year after implementation, well after I had left, several forecasters told me that they still were struggling with this complex system. As I was completing this manuscript I was informed privately that after "almost three years . . . there is still an intense dislike of the system by many [Chicago] forecasters."

As a new software program, changes, additions, upgrades, and modifications to IFPS were only to be expected. Early versions of IFPS were imperfect, riddled with awkward design features, bugs, and glitches. Some believed that these would be fixed, improving the forecast process; others were not persuaded.

The Imperatives of Organization. Like most major technological changes, an organizational imperative motivated change. For good or for ill, this change resulted from the demands of private meteorological firms, notably AccuWeather (described as "our biggest nemesis").[39] By relieving the staff of writing forecasts, they treated the provision of data and interpretation as their primary product, with the written forecasts downgraded. Optimists felt that this change would "save the service as a service" from the threats of privatization (Field notes). One confessed, "I think it actually saves our jobs. It quiets down the private sector. As long as they have this digital database, I think it provides [us] job security." Yet, a colleague commented, "Perhaps they can save on personnel costs and still have enough people available [for severe weather]. I told people jokingly that I went to Kansas City [for IFPS training] to learn the software that will put us out of business" (Interview). The system potentially allows one forecaster to manipulate the database, implying that his partner is redundant. Others suggested that the hydrometeorological technician position might be eliminated with the second forecaster responsible for that work.[40] Some even believed that this system with its national database might result in the closure of local offices, consolidated in a few regional centers.[41] Such a system would help coordinate the grids with adjoining offices, a problem that the new system magnified in that points on the boundaries of forecast areas may have very different forecasts (a problem less apparent when the forecast was for whole counties). Still others concluded that someday this system in conjunction with ever more sophisticated forecast models might entirely eliminate forecasters, making this public science an appendage of media and industry.

The Strains of Change. What did this change mean to forecasters? How revolutionary a change was it? The science didn't change, the need for forecasts and explanations didn't change at first, the personnel didn't change, and neither did the material conditions of labor. Yet, some saw that this change was as revolutionary as the creation of the nationwide network of local offices, the installation of computers, or the availability of real-time satellite or radar images. They experienced "epistemological distress,"[42] uncertain as to how work is to be done under this new knowledge regime. One anxious forecaster relied on the metaphor of *Brave New World,* with the assessment that the software was still primitive—"not ready for prime time"—a steep learning curve was required, coupled with fears of an uncertain future. I was informed that "There will be people who will retire rather than do it" (Field notes). That didn't happen in any of the offices that I examined during the process of implementation. Forecasters were told dramatically—frightened, perhaps—that "this is your future" or "for most people this will be the bulk of your career starting now.... This will be a sea-change in how you put your forecast together" (Field notes). One meteorologist even suggested (humorously, I hope) that IFPS could cause people to "become insane" (Field notes). Some forecasters—often the younger ones who relished change and were not set in their ways—were more compliant. Two forecasters phrased the matter differently and personally, "I like change. There is so much potential.... The only way you will be threatened is if you don't embrace the new technology," as opposed to the simple plaint, "I'm doing it the old-fashioned way. Dogs don't like new tricks" (Field notes). Whether changing from writing forecasts to producing those interpretations of the future by drawing lines and moving boxes constitutes a major change is not open to objective assessment but is determined by how a group and an occupation characterizes its work.

The Belvedere and Chicago offices implemented the new system differently.[43] In Belvedere, the training was rapid. Two meteorologists who had been trained in regional headquarters were responsible for training their colleagues, sitting by them and answering questions. Forecasters were expected to master the whole system at once from data input to creating gridded forecasts. Up to five days were set-aside for training, but most forecasters felt that after two intense days they could manage on their own. Perhaps because the training was concentrated and provided by respected colleagues who were enthusiastic about the system, it proceeded smoothly. If one accepted that computerized forecasting was inevitable and desirable, the Belvedere approach was advantageous, and the system was adopted easily. Of course, the culture of the Belvedere office and the identities of the workers, as was expressed many times, was that of a group that was open to change. Thus, any innovation that headquarters selected might have been readily adopted.

The Chicago office took a different approach, in part because the staff was unenthusiastic about the change and wished to delay its implementation. IFPS lacked an effective advocate in the office. Even the trainers were skeptical. The Chicago office first held what they described as a "political meeting" at which forecasters were warned about "big changes" ahead. Then training by a member of the administrative staff and a younger journeyman forecaster occurred in stages. First, forecasters used the system to create verification statistics, then they manipulated the matrices, the numerical database that produced the forecast (ignoring the geographical grids), and finally—two and a half months after the target date—forecasters began creating forecasts through manipulating the grids. If they had learned to draw their forecasts on a gridded map first, as happened in Belvedere, the other forms of training would not have been necessary and the process would have been less disruptive. But the Chicago staff felt that this process would have been too radical; they imagined that change needed to be instituted gradually. Further, rather than have a knowledgeable forecaster work intensely with each staff member for several shifts, the training occurred in bits and pieces, and familiarity took considerably longer to achieve than at Belvedere. Whenever the program developed a glitch or when the weather was complicated, forecasters were told to use the manual system for the day. In fact, one of the trainers told me that he had decided not to use the system when he was running late, commenting "I was thinking of doing [IFPS] at noon, but I didn't have it ready. Today's not the day to play hero. Today's not the day to fart around with it. There will be plenty of days" (Field notes). In Chicago's culture delay was legitimate and normative. Eventually regional headquarters sent a forecaster from a proficient office to train the staff. In time the Chicago forecasters mastered the system, but many resented the turmoil that was caused by the new forms of work.[44]

The IF-PISS Solution. Workers have strategies to demonstrate resistance to changes that they do not think will serve them or those they represent. These tactics to thwart bureaucratic demands are weapons of the weak,[45] even when those weaklings are well-paid professional meteorologists toiling within a government bureaucracy. As offices have different cultures, the resistance to IFPS was not uniform within the weather service, but such resistance emerged at the Chicago office and some other local offices. Some offices saw the new system as a milestone of meteorological progress.

IFPS is now institutionalized and forecasters have learned to live with the technology (if not always like it). But at the time of implementation, skeptics expressed resentment. Acronyms are a tradition in the weather service. Some are spoken of by their initials (the N.G.M. model, the A.F.D.), others are spoken as they are spelled (NOAA is No-ah, the AWIPS computer system is A-Whips). What to do about IFPS?

Headquarters and the Belvedere and Flowerland offices refer to the system through its initials as I.F.P.S.; however, for many—although not all—of the staff in Chicago the editorial comment implicit in speaking the initials was too prime to ignore. The system quickly became "If-piss" (and sometimes, when particularly frustrated, "G.D. If-piss")—a moniker created by one of the forecasters at the beginning of the process. The term took root and conformed nicely with the cynical microculture of the office. Staff commented, "I'll be if-pissing this afternoon" or "Bert was really if-pissed off" (Field notes).

If-piss was only one of the ways in which IFPS was colorfully designated—it exemplified "crapware." Rants aimed at the program and those at headquarters who wished to implement it were common. One forecaster claimed that because of the system he would retire the day he was fifty-five (a decade in the future) and others were convinced that younger forecasters would quit, becoming bored or frustrated with the new system, wondering, "How will I have job satisfaction for the next thirty years?" (Field notes). Blunt criticism of the technology and its makers was common. They were not involved in its creation, not sold on its value, and feared for their futures.

I never heard the term If-piss used in Belvedere and only rarely in Flowerland. There one forecaster, frustrated with the slowness of the system, commented "even the acronym sounds bad," but significantly did not mouth that acronym (Field notes).

The Autonomy of the Author and the Pride of the Scientist. A standard complaint was that IFPS was not ready for distribution, a grievance sometimes coupled with the claim that proper training was lacking. In this, complaints were consistent with other moments of technological change. These issues provoked general agreement, even among those who were convinced that the system would eventually benefit operational meteorology.

To understand the benefits and the threat, I examine claims about the long-term effects of the system, focusing on literary autonomy and scientific advances. It is not that either claim is necessarily correct, although some day consensus may emerge, but these are images that meteorologists draw from when they consider their own future and how the system will facilitate or hinder the act of forecasting.

Literary Autonomy. On my first day at the Chicago office, one of the staff took me aside to explain the new computerized forecasting system, yet to be unveiled, adding, "People feel they are losing all their individual artistry. The machine will put out all their words. They are worried about losing their words" (Field notes). A few months later I was told: "Right now we feel that we own the product, because it's our words. It won't be our words in this new system. We'll be putting out numbers. *We're not authors anymore*. It's not the same as putting out a product that everyone reads. We're *just* putting out numbers and the computer will create the words. It won't be the same" (Field notes, emphasis added). Another claimed, "Now with IFPS it's the computer writing *my words*. When I started, coming up with the words was crucial. Now the computer does it" (Field notes, emphasis added). I detailed the importance of words to forecasters, but under the IFPS regime, it is a *machine* that steals their soul. Yet, it is not just the words, but these words constitute a *narrative*. As a Belvedere forecaster told me, "I like to look at the data and tell my own story, and that's been taken away from me" (Field notes). I felt that I was watching professional chess players facing the obliteration of their expertise under the pressure of Deep Blue.

The frustration became intense at moments in which the computer prevented the forecaster from selecting the words by which the forecast would be known. One put it bluntly, "It's just a tool, but I'm the forecaster!" (Field notes). They were blocked by technology, just as some computer users feel when a program informs them that they are not *permitted* to save a file. Who made the chips our cops? This is a battle between man and machine, as in the claim, "IFPS puts out what it wants to, not the wording that is important to us as a human being. IFPS doesn't know that is important from a human perspective" (Field notes).

Recall the discussion of the use of "fair" as a meteorological term: "I did a forecast of 'fair,' and they don't have 'fair' in [IFPS]. I have to do some other malarkey. It's 'fair.' It's not 'clear.' I'll make it 'clear.' No, I'll make it 'scattered clouds.' They're frowning on 'clear,' what the Hell?" (Field notes). Under the IFPS regime "fair" could only be included in the forecast if the meteorologist edited the machine-generated text. These forecasters must create work-arounds and make-dos, evading the technology and maintaining the authority to decide. One forecaster made this explicit through his desire to create wording, whatever the database might be, defeating the aim of the new system: "There will be a day when I won't be concerned about the wording. Maybe that will be after I retire. At some point we will only be dealing with the gridded database. Let's see what laughable zones come up. . . . Now, I'll go in and make the changes to have it come out the way I want it. They wouldn't like it if they knew, but that's the way we do it" (Field notes).

Eventually manipulating the gridded maps to produce desired words became easier, but it was still the words that counted for the occupational self. The implicit threat is that under the new regime the public will not notice the change. As one administrator put it skeptically, "A lot of people feel ownership of their words. . . . It's something the forecasters worry about, but nobody else does, even their family doesn't care" (Field notes).

Scientific Advances. Perhaps this account has painted IFPS in shades of dark gray, a function of the people whom I came to know and respect, but their views, while not rare, are not universal. As noted, about half the National Weather Service personnel believe that the new system will be beneficial. For them, greater attention to meteorological detail is a blessing. A meteorologist now no longer needs to forecast for an entire county but can consider subtle topographical features. The weather in Chicago can vary substantially from the lakefront to the prairie suburbs at the southwestern edge of Cook County.

This system permitting attention to detail can enrich scientific claims. The challenge is intellectual as one meteorologist noted, "You're going to have to think at a faster pace." Or as a forecaster said of his partner after his first day using IFPS, "I smell burning wires [brain synapses] from there." The primacy of scientific thought is evident:

> I see the benefits from it. Just putting out the words is too narrow. . . . [Eventually] you'll concentrate on the meteorology, on what you think will happen. You let the machine come up with

> the words. You can still manage words [through editing]. In some sense, it's more fun. It forces you to change how you think about the forecast. You still get the picture in your mind, and you convey that in the graphics and the words flow from there. (Field notes)

: : :

> You can get a much better reality view of what is going on over the area. . . . With a gridded database you can show that trend. You can even show a lake breeze because of the smaller grids. . . . It holds a lot of promise in terms of how we present the weather in terms of what we can say. . . . If we can save time in the composition of the products, then we will have more time for more scientific assessment. (Field notes)

These workers privilege the idea of scientific interpretation over the loss of autonomy in the creation of texts. Is it the science to be communicated or is it the story? They are interconnected, but the relative weight chosen impacts one's attitude to change. Yet, these forecasters recognize that someone else (i.e., headquarters) has made a decision for them that forces them into a particular mode of scientist rather than the hybrid public scientist that has governed their worklife and that they have come to imagine is the way that their work ought to be.

Ultimately the change becomes the tacit reality, and recalling how work was constituted prior to the change is difficult. No doubt this is increasingly true with this technology. Meteorologists have always been, for better and for worse, on the cutting edge of technologies of data gathering and dissemination. IFPS is part of a historical set of transformations that shift the authority of the forecaster from being the font of wisdom to being a data manager. The case of the boxes that are used, critiqued, loved, and sweated over by the staff of the Storm Prediction Center reminds us that the concerns that Chicago forecasters have about their words can also apply to other forms of communication. As anxious as these forecasters were during my research, they may become as protective of their lines as they now are of their words. They can become artists as they were poets. Still, a meteorology that didn't use words to communicate is difficult to imagine. The question of who constructs the words, who has the right to alter them, and who ultimately controls them as they enter collective knowledge remains to be determined. The issue ultimately is not words, but control.

Writing on the Winds

When we glance at a forecast on the Weather Channel, it may take effort to recall that those words are not merely instrumental but are expressive products. Words are not simply shared but are owned. They are integral to the doing of science. Whereas much social science analysis has emphasized the rhetorical qualities of scientific reports, notably journal articles, the worklives of operational meteorologists demonstrate that even a line of formulaic text can be pregnant with meaning and emotion. Words leave themselves open to standardization both within a local community and through decisions of a hierarchy. Mark Twain was right to claim that weather is a literary speciality. While the public treats these modest texts as transparent, they may be subtly, unknowingly influenced by how words—fair, clear, partly cloudy, mostly sunny, sprinkles, showers, flurries—convey different moods. Do partly cloudy and partly sunny mean the same thing? Do they give the same emotional picture to an audience, and, if not, which is preferable: should the sky be half full or half empty? Should forecasters be wily optimists or sly pessimists? Would you rather be pleasantly or unpleasantly surprised?

Words matter in this public science, for words are the means through which we communicate ideas, reveal identity, and demonstrate authority. Other technologies are sometimes used, particularly by media forecasters, such as boxes, pictograms, colorful charts, maps with squiggles, cartoons, or computerized images,[46] but even here, although words are pushed to the side, the control of communication is central. Forecasters desire to maintain the authority to inform the public as they think proper, revealing their skills.

Any system, such as computerized forecasting, that threatens this control is troublesome, unless forecasters can convince themselves that they benefit: in this case, with the potential of expanded scientific authority. Not only do individuals differ on these effects, but groups, offices, and communities differ as well, generating spicy debate over these occupational changes. An ethnography of these same offices today, a few years after the implementation of this system, would demonstrate how this system has become routinized and naturalized as part of how forecasters do their jobs. Given the march of technology, other changes will impact the identity that forecasters struggle with as individuals and in offices.

The opening theme of this chapter reflects the centrality of collaboration, both internal and external. No individual can create scientific knowledge without relying on a team and on equipment. A forecast is a relay race in which predictions are passed off; each forecast shapes future ones, modified as new knowledge is available.

Not only does a local office require coordination, but connections among offices also build on negotiation. As each County Warning Area abuts several others, forecasts should be smoothed out so that weather claims flow as naturally as the fluid dynamics of the air. Through talk and documents, this smoothing occurs or is ignored. As in the case of the Storm Prediction Center, when large blocks of the nation must be organized in the face of groups with varying interests, the challenge of creating a happy negotiation is real. Meteorologists do not always work together as effectively as they might, but in a system that depends on *integrated autonomy* that goal is to be devoutly wished.

5 Ground Truth

"We're going to verify our [heat] warning with a body count. You've got to verify somehow. . . . You verify twice. Once by the heat and then by the number of people killed."

Field notes

: : :

What does it mean to be "right"? Much scientific work depends upon communal criteria by which claims can be assessed as correct.[1] But how are these criteria used in practice? It is one matter to predict, quite another to evaluate a prediction. This is a problem of applied epistemology. While I focus on a single occupational domain, that of operational meteorology, this analysis applies to other scientific occupations and beyond. What is true in the case of public science pertains in part to other work domains in which a logic of accountability is demanded.[2] What I claim about meteorology applies to truth claims propounded by economists, physicians, and political pollsters. For institutional legitimacy technologies of verification must be established. Once established, these organizational demands for accountability must be instituted (and evaded) by those workers who are to be evaluated.

I begin by focusing on how warnings of severe weather are evaluated after the fact. These warnings cause people to act and, when missed, can claim lives. But how does one know what really happened? And how does an organization establish a system by which these judgments become routine? How can this system be gamed by workers? Put another way how are the technologies of truth organized within a bureaucracy?[3] I examine the case of a deadly tornado that touched down in La Plata, Maryland, on April 28, 2002, during the period that I was observing at the Storm Prediction Center in Norman, Oklahoma.[4] My focus is to analyze how creating forecasts (both mundane forecasts and watches issued by the SPC) and keeping weather records result from extrameteorological features. An appropriate forecast is shaped by social needs, as are those records that are selectively kept of weather history. Focusing on how predictions are judged to be suitable claims, I describe *ground truth* in its various forms. While ground truth is a term used by meteorologists, the underlying problem is known by all those who must assess a reality to which they do not have direct sensory access.

We can speak of three models of judging the accuracy of a claim: *experimental*, *observational*, and *predictive*. Each provides legitimacy within particular knowledge regimes, as how and what one knows is tied to accepted methodologies. For an experimental scientist comparative accuracy is sufficient; absolute truth is not crucial.[5] So long as bias or error is equally distributed across conditions, the experiment is sound. The goal of experimental research is to determine whether one condition has produced significantly more or less of a dependent variable than another, given statistical criteria. Typically the precise level of that dependent variable is not crucial, since the comparison is of two artificial conditions.

Observational scientists have a different set of concerns. For them, the precision of description is crucial. If their observations do not fall into a narrow range of collective expectations, the outcome is not judged to be credible. But how is this zone known? The problem is to judge a real reading from which the observed reading meets or deviates. Ultimately one must accept these claims on faith, comparing the results of similar readings by others, readings through different measuring instruments, and the reputational regard in which the observer is held. The judgment of accuracy is communal. The third type of accuracy, the one at issue here, is that found in much public science and is predictive. Workers forecast events, and then these claims are checked for accuracy as the future unfurls. The test is how successful are practitioners in forecasting. These tests are found in petrology, climatology, medicine, and meteorology. Errors, when wide enough, can spark legal liability (e.g., malpractice or negligence), needless expense, and mass inconvenience.

Of the three modes of accuracy, it is only the third in which a forecast of a scientist can be compared with "objective" measures of what has been forecast. These measures are treated as separate from personal judgments of practitioners. We believe that we can determine the relationship between the prediction and the subsequent event, allowing us to determine the accuracy of a claim.

Of course, not every claim is verified. Some are not deemed sufficiently important to be tested, and this selection of claims produces bias in judging the accuracy rate. In the case of severe weather, for instance, a forecast of a tornado would need to be verified to determine whether the event did or did not occur. Meteorologists may decide, however, not to issue a warning for a tornado—itself, in effect, a forecast. The absence of a forecast is only noticed if a tornado occurred; otherwise it is not a forecast. A correct belief that severe weather will not occur does not count toward organizational effectiveness. This bias in which only selected predictions are tested is labeled "the forecaster's dilemma."[6]

Meteorologists are an apt case for analyzing the problem of judging accuracy in that as a matter of occupational routine forecasters produce claims about future events that are checked and evaluated. Their status, both as individual workers and as a group, results from how well they predicted. Did the weather occur as expected, and how can one judge if that is the case? Further, as federal employees, they are constrained by the logic of accountability, making verification institutionally central.

Meteorology is both typical of scientific knowledge and a special case. The field is distinctive in that the public assumes that weather patterns are easily knowable. Anyone—with the proper instruments and the ability to read those instruments—should, from this perspective, be able to describe the weather and determine whether the forecast is correct.

Forecasters do not require such esoteric devices as electron microscopes or cloud chambers. Thermometers, rain gauges, and barometers are user-friendly, and even specialized equipment such as satellites, upper-air balloons, and radars can be appreciated by the general public. Weather is experienced by, if not precisely measurable by, our sensate skin. Weather appears easily knowable in a way that the claims of cosmology, microbiology, and particle physics are not. Further, the events of meteorology transpire in real time, unfurling on a human scale of knowing.[7]

My informants speak of the challenge of knowing what occurred as constituting what they term the problems of *ground truth* and *verification*. The former, referring to the reality of meteorological conditions, is used when a forecaster is attempting to determine what is happening in a short-term severe weather event, while the latter covers the organizational assessment of both severe weather and longer-term forecasts.

On the Ground

Sitting in a comfortable, secure office, forecasters are required to judge dynamic and tempestuous conditions. Surrounding them are computer screens on which are displayed the output of measurement devices and modeling systems. These images are supposed to tell forecasters what they *need* to know about the world that they have the responsibility to divine.

Although the machine readouts are often accepted without question, this is not inevitable.[8] The mechanical claims must fit what workers believe to be true of the atmosphere. They must be consistent with the commonsense assumptions of the trained professional. Consider a case from the Belvedere office:

> A little after 14:16 Zulu, or Greenwich Mean time (9:16 a.m. in Belvedere), the equipment measures a brief temperature spike for Midwest City to 60 degrees, although the high temperature both before and after was about 55 degrees. Garth and Dom discuss whether this temperature is "real" or whether it is an equipment error. Since it is a reading from Midwest Airport, they speculate that it might be jet exhaust, although this problem has never happened previously. They notice that the temperature in North Boyer was 60 degrees at 14:23, which made sense because the clouds over North Boyer were breaking up, and perhaps explained the temperature spike in Midwest City. However, they soon realize that the 14:23 time was from their 24-hour clock, representing 2:23 p.m. They puzzle over the actual temperature in Midwest City, but accept that the North Boyer temperature is correct in that the clouds justify that temperature. Dom explains the Midwest City temperature: "No one on site could tell you that the observation was accurate. There is nothing meteorologically sound that said that this did or did not happen. There is no way of knowing it." They decide that the North Boyer temperature reading is correct, and the Midwest City temperature is incorrect. Garth comments about automatic equipment replacing observers, "Now you have greater detail, more data, but you have more doubt when it is anomalous, because there is no one there to interrogate the data." (Field notes)

To know a meteorological reality, these forecasters rely on their shared assumptions and experiences of how weather systems operate. If one eliminates human observers, one also eliminates the possibility of discretionary questioning of the data, other than by challenging whether the machines are working properly.

New forms of data collection pose, at first, a similar problem:

> Dom explains that when they went from human judgment to laser measurements of cloud heights, the heights increased substantially. He adds, "The cloud heights certainly look lower than that. It took about six months before we believed it." The mechanical claim was not proven, but workers were worn down by the consistency of the machines. (Field notes)

The search for *ground truth* is focused on deciding what is real, given organizational demands. Forecasters find themselves hostage to claims that rest outside their own expertise, and, so, their willingness to criticize these readings represents an act of professional heroism.

In contrast, one form of ground truth stems from experience. When weather is known personally, it is embraced. On a day in which a strong possibility of flooding existed in one river basin, forecasters issued a warning because Greg, who lives in the area, reported that most of the streams he saw while driving to work were over their banks. This was enough lived experience for the forecaster on duty.

The problem of discerning ground truth is more acute at the Storm Prediction Center. Much of their information about storm damage is filtered through local offices. When severe weather strikes, these offices may be too busy to issue detailed storm reports. One forecaster at the SPC confessed that they are often uncertain of what is happening: "We'll see the storm on radar, and we'll hear that there is storm damage, and then we'll go home [and turn on the television to learn what really happened]." He describes a major tornado in 1998: "We heard that a tornado had hit the west side of Birmingham [Alabama]. The death toll just kept on climbing. We didn't know until noon the next day. We usually don't hear. We don't hear about significant happenings. We don't want to step on people's toes [by calling]" (Field notes). Staff are frustrated by the inability to receive a timely assessment of their forecasts. They "see" what is happening on the radar, but their eyes are limited by the mechanical. Sometimes the first indication of severe damage is a call from the media.

This blindness was evident on April 28, 2002, the day of a major tornado outbreak across the United States, eventually leading to a deadly tornado that hit La Plata, Maryland (discussed in detail below).[9] This was a powerful storm, but learning what was happening was a problem:

> During the middle of the day Rod comments, "There's some mean looking stuff. I haven't seen radar that looks like this in a long time." However, he admits that he doesn't know if these supercells are producing tornadoes. The radar only indicates a visual representation of vorticity. The beam is aimed too high to determine whether tornadoes have formed. Linc points out that "We might not know until we get damage reports." Local offices are supposed to notify them of severe weather through Local Storm Reports, but LSRs receive lower priority than issuing warnings. Guy comments, "We're held hostage to those LSRs. It's not usual policy for us to call. That makes it kind of hard to know how well we did. . . . Usually when the very worst things happen, there is silence. There's a lot more things that they worry about than sending out an LSR to us." Linc notes that they received a report of softball size (3-inch) hail: "It's surprising that there are no tornado reports, given the hail." Gordon remarks, "We don't know what's going on." Rod adds, "With all those warnings, there is not so much concern with ground truth. They are probably too busy." (Field notes)

Later a strong storm cell heads into Maryland, the storm that eventually produces the deadly tornado, and Dean, looking at the radar image, "That's nasty looking. Holy Cow. . . . That thing is really ugly looking. It's really scary." They see evidence of debris *on the radar image*, but still haven't received a call. Guy sighs, "The lack of news from that is disturbing." Only later do they learn the extent of the damage in suburban Washington (Field notes).

This episode emphasizes the difficulty of translating what is evident on machines to claims about the state of the atmosphere. As an institutional matter, once the SPC has issued their watch, their job is over. Picking up the pieces is the responsibility of others.

Verification Games

Surely it is reasonable to ask of any activity whether it meets the goals that participants have set for themselves and that others have set. How are we to assess work? Here the focus shifts from forecasts to what we might label *hindcasts*. That organizations have investments in outcomes makes decisions of how to evaluate those outcomes social and political as well as technical and scientific.

When a government forecaster makes a prediction, citizens have a reasonable expectation that there will be accountability, and this accountability shifts organizational priorities. Yet what does a claim that a forecaster is *accurate* mean? Are we substituting precision for validity?[10] John King's observation about policy arguments is pithy and true: "some numbers beat no numbers every time."[11]

Consider the following claims:

> Accuracy improved but little in the first half century of the weather service, hovering around the 75 to 85 percent mark—and this was for the next day's weather.... In 1941, five-day Weather Bureau forecasts had an accuracy rate of only 48 percent for temperatures and 16 percent for precipitation.[12]

: : :

> A three-day forecast in many areas of the country is only about 50 percent accurate.... It means that the forecaster has "blown the forecast" three days into the future just about as many times as he's gotten it right![13]

These passages are from books by science writers sympathetic to operational meteorology. Yet, in neither case do the authors indicate what accuracy means. The numbers sound precise, but precise for what? What does it mean for a forecast to be 50 percent accurate or even accurate 50 percent of the time? Consider temperatures.[14] If a forecaster predicts that the temperature will reach 54 degrees and the high is 53 degrees, is it wrong (current thermometers are reliable to plus or minus two degrees)? Suppose the high is 51, 45, 36. At what point is the temperature forecast wrong? In Chicago's Cook County the temperature by the lakefront may be 42 degrees and the temperature in the southwest corner of the county may be 64 degrees, a range that is by no means impossible. Averaging these two readings is 53 degrees. Is the forecast correct? Are we to measure high temperatures, low temperatures, or average temperatures? If average temperatures, how should this be measured? The official Chicago reading is from O'Hare International Airport. Should that be the measuring point, even though it is far from the population center of Cook County? Do we average the high and low temperature or average the reading for each hour. These are organizational choices.

Judging precipitation is as problematic. Suppose a forecaster claims that there is a 40 percent chance of rain. Is the forecaster correct if it rains or if it doesn't? Suppose there are occasional or scattered showers. Should O'Hare serve as the measuring point for rain or snow, as the organizational practices require? It may not rain at O'Hare but may in downtown Chicago. So, what does accuracy mean? When one moves from temperature and precipitation to the accuracy of a forecast as a whole, things become even more complex. As weather consumers, we *sense* when a forecast is *right*, but this refers to whether the forecast inconveniences us (getting us cold, hot, or wet, or needlessly canceling our plans) and is not linked directly to numerical precision.

Given these challenges, verification is construed by the NWS as an organizational practice. As a government agency at a moment in which accountability is a watchword,[15] the ability to claim institutional effectiveness is critical. The head of the National Weather Service at the time of this research, retired general Jack Kelly, has been successful bureaucratically by emphasizing "objective" (or at least numerical) indications of organizational competence.

That the bureaucratic emphasis on numerical indicators of effectiveness and using them for organizational display has not always characterized the National Weather Service is suggested by the claim of an administrator, "I was a manager who believed in verification before verification became cool" (Interview). A similar view is evident in the claim of a forecaster from an earlier period, "We don't keep track of our accuracy rates and compare our scores with the other stations. That is childish. We're all using the same materials. Sooner or later you'll be right and they'll be wrong and vice versa. When you're wrong you don't want to commit suicide. It's really not that big of a deal."[16] Such a claim would no longer be spoken publicly. Today, performance-based measures are demanded, so the question is how staff actions can be constructed to demonstrate success. Analogous to evidence-based medicine or accountability in education, we find performance-based meteorology.

Crafting measures of organizational competency is a bureaucratic strategy, and a successful one in that President George W. Bush and his aides, as well as congressional leaders, cite the NWS as an agency that works. They demonstrate their confidence by pointing to the statistics that the agency has compiled from their bureaucratic self-interest. It is not that the statistics are false, but the issue is analogous to teachers teaching to the test, rather than measuring learning.

Organizational impression management operates both in central administration and at the outposts of the organization. In this, headquarters establishes procedures to measure effectiveness. The dilemma is that the administration has two goals: that the local office follow the procedures precisely and that the local office demonstrate that they are succeeding in their tasks. Ideally the two goals fit together, but what happens when following assessment policies leads to a realization that goals are not being met? One hopes that the organization addresses the failure, but often the system is gamed, demonstrating that, whatever is happening, the data can be massaged so that progress appears. Weather service offices provide a nice instance of this, but it is also found in fast-food restaurants, university departments, park service offices, police stations, and other domains in which a central authority desires that local units compile information for their own evaluation.

As with any organizational practice, the doing of verification is complex. Local offices are fundamentally concerned, however, with two forms of verification—those assessing routine forecasts (temperature and precipitation) and those assessing severe weather warnings.[17]

For every public forecast (two per day), forecasters fill out a Coded Cities Forecast (CCF), in which they specify the high and low temperature (for a period of a week), and indicate the probability of precipitation. Each office has a set of observing locations for which they make internal forecasts. In Chicago, meteorologists forecast for O'Hare and Rockford airports. The test is whether a forecaster can best the predictions made by statistical models (what are called model output statistics or MOS).[18] The organizational goal is to beat MOS.[19] When the models are substantially wrong on forecasted temperature, they are easy for a skilled meteorologist to beat. But even when the models are effective, strategies permit forecasters to beat them. If the model suggests a 90 percent chance of rain, a clever forecaster can best the model by specifying a 100 percent chance of rain. Assuming that it rains, the forecaster will get credit in the verification statistics as having greater confidence than the model prediction. Precipitation verification compares the percentage likelihood of precipitation in light of what happened at the recording station.

The verification of severe weather operates differently. Once severe weather is forecast, the test is whether the predicted event occurred. Was there a tornado or did a thunderstorm reach severe criteria (58 mile per hour winds or 3/4-inch hail)? How can one judge whether a forecast was correct, since images on the radar are not sufficient to verify? A human observer is necessary unless the storm passes directly over a wind gauge. On occasion, storms are verified after the fact through damage to trees or property. If forecasters learn that a tornado or severe thunderstorm occurred that they haven't predicted (and needless to say, searching for such a storm would not be a priority), this counts against their verification statistics, because the NWS counts the number of severe storms for which no warnings were issued.

Inscribing Ground Truth. In verifying forecasts, meteorologists are at the mercy of their machines. The automatic temperature-recording machine is considered accurate to within two degrees. Measuring snow depends on whether one considers the water content of snow or the degree to which it is packed.[20] The height at which wind is measured affects wind speed. Even small changes in where equipment is sited can noticeably alter the data. As one forecaster explained:

> I worked at the airport. Sensors ideally were supposed to be placed close to midfield, representative of the airfield. We were then losing the temperature and dewpoint there, and they found out it was because the cable had been cut and spliced so many times water was getting in, kind of shorting it out. So they just moved the equipment, the temperature-dewpoint sensor, closer to where the office was instead of being out on the center of the airfield where it was level . . . So it dropped off maybe 12, 15 feet. (Interview)

When equipment is replaced or is moved, the continuity of records becomes confounded, and the meaning of the measurements becomes problematic.

Verification and Value. Verification is tied to a system of values given institutional priority. Whatever the forecasters might wish to use to determine their successes (and failures) in prediction, the agency for which they work sets procedures that they must follow—or evade as best they can. In this meteorologists face a tension between professional autonomy and organizational constraints similar to that faced by scientists who work for industry.[21]

Although the National Weather Service does not currently use verification statistics in setting salaries for individual forecasters, offices are required to gather, and to display, their collective verification statistics, transforming these numbers into a means by which reputations are established. Regional and national administrators treat these numbers as an "objective" basis on which to distribute resources to successful offices. Verification statistics are taken as *signals*, among the few systematic signals available to permit a hierarchical ranking of effectiveness.

The Belvedere office, with a culture that focuses on accountability, emphasizes their statistics by placing charts in a public area. Staff claim that the emphasis on accountability is beneficial. As one told me about verification statistics, "It is a good check on our mind. If there was no verification, I would still forecast as I did, but it's still important to know" (Field notes).

A paradox is evident. Meteorologists realize that often enough forecasts do not verify and that forecasting with verification in mind may make one overly cautious, ignoring possible, but unlikely, risks. Yet, this is the only measure of competence available. Forecasters are exhorted not to worry about verification—to do the best they can—but they are then faced with the numerical representation of what is ostensibly a secondary concern. Verification statistics, however secondary they might be, are a rhetorical resource used by administrators. One urged his forecasters to be concerned with the science, but simultaneously that "We're constantly watching our performance measures. . . . We've been top gun for winter weather" (Field notes), revealing the multiple, cross-cutting concerns involved.

Others are rhetorically far more suspicious of verification, although they too bow to the organizational demands for a metric accounting of their abilities. The basic complaint is that verification is a game and doesn't provide a "true" measure of forecasting. Rather than seeing how human and model combine (the "man-machine mix" or "how the forecaster interprets the model against what actually happens" [Interview]), the verification of daily weather involves comparing the human forecast against a machine-generated model. The organizational question is whether a human forecaster improves the predictions, but for the public, the question is different: Is the prediction seen as useful? As one meteorologist explained:

> I want to know what I've done compared to reality, because my ultimate goal in the perfect world is to put out the perfect forecast, which I know is an unrealistic goal, but I think the effort should always be striving in that direction. Higher ups in the Weather Service compare us more to the numerical models with the idea that if more of the numerical models are better than a forecaster then what do we need a forecaster for? . . . With a lot of this fair weather that we've had this winter, the models do as good as we do, maybe at times a little better. For fair weather, we're not as cost-effective perhaps. But with this big winter storm coming up, I can almost guarantee that time in and time out,

> we will have a tendency to beat the models. You don't pay the fireman for when he's sitting in at the fire station waiting for the bell to ring. You pay him for when he's at your house putting the fire out. When it really matters to the public, are we doing better than the models, and if the answer is yes, then I think we're doing our job. (Interview)

The image of verification as an organizational game is routine,[22] even among those who are resigned to the practice. I was told that "everyone plays the system," playing "a numbers game" (Field notes): "They're playing numbers against the models. They're not forecasting . . . they say, 'Well, if the numbers say this, so that's what I'm going to say, so my numbers match.' . . . [They're trying to] beat the numbers. And I think that is a terrible pitfall that we can fall into very easily. And I've seen forecasters here do that." (Interview). An administrator, noting that the game involves impression management, instructed his staff, "We're supposed to post these things [verification statistics], so find a place where we can post them so no one will see them, unless we're doing better than average" (Field notes). He considered these procedures little more than "schemes" that were easily manipulated, involving "cheating" to succeed.[23] I never learned if they were doing better than average, but the sheets were posted in an obscure location.

One administrator commented, half-joking, that perhaps they should not attempt to warn for small F0 tornadoes, since "You're doomed to failure as far as the numbers. . . . We just don't get the verification on F0 tornadoes." Indeed, of the 282 tornado warnings in this CWA over fifteen years, only 121 were verified (43 percent), most often through hail or wind damage; only 38 (13 percent) were verified by visual sightings. The same was true of severe thunderstorms. Of 1924 warnings, only 798 (41 percent) were verified. The number of "actual" tornadoes or severe thunderstorms is anyone's guess. Presumably, at least some of the false alarms were correct warnings and some verified storms were accepted on the claims of mistaken eyewitnesses. One forecaster suggests wistfully, "Well, our numbers will go up if we have a year in which there is a bunch of F2 tornadoes. . . . It will help our statistics" (Field notes). Such large tornadoes are far more likely to be seen and known through their destructive power. Warning, even for a real event, may have negative organizational consequences if evidence is unavailable to support the claim. A distinction is drawn between what is important and what is organizationally necessary.

Creating Verification. The mission of any science is to present the contours of the "real world" in a way that audiences accept. Meteorology is no exception. As a result, having an organizational gauge of how successful one is in predicting severe weather is crucial, even if, as described above, most warnings do not verify. But how does one know what has happened, given the absence of data? How can meteorologists discover ground truth?

A tornado is real enough if one is caught in its midst, but most tornadoes are not much stronger than a heavy wind. If the vortex skips over a field or lake it may leave no marks. Sparsely populated land tells no tales.[24] If the timing is wrong—when the world sleeps—in effect, the storm never happened. The same is true even if the storm was observed but an observer doesn't call. If the public doesn't respond, only the physical impact of the storm remains: holes in a tilled garden from hail, broken branches from high wind, or a circular pattern of debris from a tornado. Looking at a small storm that produced 1-inch hail, a forecaster commented, "Those storms don't look like anything on radar. Those [hail]stones must have just fallen where someone was looking because they don't look like a big area" (Field notes). In such cases, there is a strong temptation not to warn for storms that might barely meet the criteria for severe weather. As one forecaster explained, "What we had was really pulse storms. Even when you warned it would have weakened by the time the warning went up. My preference is not to warn because it is on the border. The chance of getting a sighting would be rare" (Field notes). Whether this is desirable for public safety, it is justified as an organizational strategy to maximize verification. In a sense the professional obligation to "get it right" and to protect the public stands opposed to the organizational demand to demonstrate that you got it right. The pressure to perform well in one sense threatens it in another.

Other forecasters detail the challenge of determining the true weather conditions in the face of an opaque reality and a public that doesn't always report what it knows:

> We say we can use radar to initiate a forecast for a tornado. We cannot use the radar to verify. A lot of these things happened in open areas where there is no opportunity to get a verification. So it happens to people where they have not issued because they know they can't verify[25]... Burke County had a severe thunderstorm going through there, and one of the forecasters said straight out, "I'm not going to put out a warning for that

> because nobody's ever reported anything from there, and I've never been able to get a verification." (Interview)
>
> : : :
>
> I used to do storm data which is the verification tool for severe weather, and particularly on small tornadoes you have no idea. . . . You make a decision based on a [tiny] newspaper article and you sit there and you say hmm, I wonder if that really was there. Or you grab onto a word or a group of words in there that convinces you that it must have been a tornado and that's how it's verified. For maybe F3 and above [tornadoes causing "severe damage" according to the Fujita scale], maybe the verification is pretty good. Below that it's a joke, it's not realistic. (Interview)
>
> : : :
>
> Not everything gets reported. If you see your tree blow down in front of your yard, and you don't happen to report it into the police, and they don't happen to report it in to us, then I don't hear about it. So then it's like we missed the warning, but it wasn't missing the warning. So, if farmer Joe was there, he doesn't report "I had 3/4-inch hail and part of my cornfield was flattened." Well, if he doesn't pick up the phone . . . even if he picked up the phone and reported it to somebody and that somebody now doesn't report it back here, so if we don't hear about it, then it didn't occur. But did it occur? Of course it occurred. Just because we didn't get a report doesn't mean it didn't occur. (Interview)

The absence of a report is not the absence of a storm. Even if a report is made, a decision must be made as to how it is to be treated. Forecasters want reports, but not if the reports do not support their claims. For weather spotters to be organizationally useful, they must relate what forecasters want to hear. Forecasters desire public responses, but calls from the public can be a nuisance, providing incorrect information, distracting from required tasks, or contributing to forecasting errors.

The Public as Problem. The verification process assumes a direct relation between what is claimed and what actually happened, but this is problematic even when dealing with eyewitness testimony. These claims are often from individuals who have not been trained, those observing weather conditions at a distance and/or under conditions of poor visibility, and those in situations in which excitement or fear may cloud judgment.

To create a corps of trained observers, local offices arrange several dozen "spotter" talks each spring, training employees from public agencies and the interested public to watch for severe weather. The NWS hopes to recruit a group of amateurs to aid in meeting organizational goals, but in the process they create the problem of incorporating amateur knowledge within a professional organization.[26] These members of the public and of public service organizations may not use the same criteria of interpretation and reporting that professionals do. While the weather service wants to extend their eyes and ears through a corps of spotters, permitting them better to warn of immediate threats, such a structure may incorporate incorrect or misleading information into the decision-making process. Garbage in, garbage out.

In the two hours of the training session only rudimentary material can be covered and there is no test given to assess who has learned the material. As a result, despite the seeming value of having additional reporters, attitudes toward spotters are ambivalent. One forecaster estimated that only one in twenty will become useful. As a result, forecasters may take a jaundiced view of public reports:

> We don't want the public to report. The vast number of people have no idea what they're looking at. . . . You've got to rely on the so-called reliable people [police, fire or emergency management workers] . . . Most people are basically deadwood. [He refers to a tornado report in an outlying county] You see how some people panic over nothing. They just let their imagination run wild. . . . The public is going to become totally useless now that they all have cell phones in the cars. They see everything. . . . That's why we're paid big bucks. Not to panic. Here's a weather law. If it don't spin, don't turn it in. When I go to a spotter talk, I hammer that in. (Field notes)
>
> : : :
>
> I never put out [a warning] unless I see it on the screen. [Public reports are] bogus. I ask did you see it. Anyone who picks up the phone and doesn't question the call is not doing a service. I tell people at spotter training that I will question them." (Field notes)

In practice, spotter reports are given variable weight on the basis of reputation and the assumption of competence. Observing a weather service office, Keli Tarp[27] reported:

> A spotter gave a report on amateur radio of a tornado from a storm that did not look particularly threatening. When the meteorologist who took the report told the warning forecaster, he responded, "I don't believe that." The first meteorologist then said, "That was X, I would believe it," and the short-term forecaster concurred, "I don't think we have a better spotter." The warning forecaster immediately began issuing a tornado warning for that county, even as he was saying, "It doesn't look like a tornadic storm."

As Tarp noted, spotters gain reputations that influence how their reports are treated, part of the way that all scientific claims are evaluated.[28] Such reputations are also tied to their organizational role. Calls from "Joe Public" are treated with less respect than those with recognized institutional positions. When the public calls, the report is likely to be accepted if it supports the forecaster's idea or if it verifies a warning, but it may be dismissed in the absence of other human or mechanical reports if it claims something that a forecaster doubts, if it doesn't match other claims, or if it means that the forecaster hasn't verified a severe storm:

> Randy informs Bert about an unwarned thunderstorm, noting that the County Emergency Management office "has a [public] report of ping-pong ball size hail, but they have two spotters up there who didn't see anything." Because the report would have meant that they missed severe weather and because it was an indirect public report, the forecasters don't take it seriously. Burt jokes, "Right place at the right time. . . . We got calls from the media, 'What did happen?'" Randy says, "The county heard on the radio that they had ping-pong size hail, but they think not." They treat the storm as though it was not severe. (Field notes)

Institutional standing provides the presumption of legitimacy, and, as a result, in contrast to the variable acceptance of public reports, those whose role provides institutional standing are given deference. The validity of a report is a situated evaluation, based on organizational credibility and authority.

Manufacturing Verification. Given the demand for forecasters to verify their predictions, strategies are developed to massage and generate verification. The local office verifies its own reports, so, as a meteorologist put it, "it's like the fox guarding the chicken coop." Another claimed, "You prepare the man to cheat." A third commented about the verification of flood warnings, "Since we're the ones who put the data on the national system for verification, we can take poetic licence" (Field notes). Is an impartial verification system possible when evidence is ambiguous and subjective? How can validity be generated when the evaluator is not a disinterested party? Because the verification of severe weather is not checked, it is easy for forecasters to claim that the public verified each of their warnings: a public of thin air. I did not see this, but suspicion was expressed when other offices verify every warning. One forecaster comments: "I hate to say it, but I think they may have made it up." Forecasters search for what is in their interest to believe:

> The Chicago office is in the middle of a severe weather outbreak, having issued a tornado warning.
>
> LEWIS: I need a severe hail report from McKinley County, if anyone's going up there.
>
> SID: Spotter says half-inch size hail, east side of Medford. [They need 3/4 inch to verify.]
>
> RITCHIE: We couldn't get a quarter inch when we need it.
>
> SID: That's what the man said. You can influence what I write down.
>
> Ritchie calls Medford to see if he can verify the thunderstorm. They issue a tornado warning for Shelton County, but do not receive confirming reports from spotters: "All these spotters, and we don't hear squat back from any of them." They finally receive a call from a spotter in Shelton County who sees dime size hail and a green sky [indicating a tornado]. Sid reports, "No sign of a tornado, but then he's in the basement. Smart guy."
>
> SID: There's a citizen report of a funnel over Afton.
>
> LEWIS: That's a good place for it to be [given the radar image].
>
> SID: I'll put it down on the form, but funnel clouds don't count [to verify tornadoes].
>
> They receive word of a waterspout in Lake Michigan near Afton. Sid remarks, "We know this is our tornado. It wasn't a waterspout, it was a tornado that hit the water." He decides that it was a tornado that hit the beach and then went into the water, and he tells others, "We had a confirmed touchdown right along

> the beach, then it went off-shore. Or it was right off-shore, because we had a spotter who watched it right off-shore. We had really good rotation." Although the spotter says that it was in Shorewood, they decide that it was actually in Afton, adding, "That makes more sense with what the spotter saw."
>
> Bert says about another tornado report, "Two residents say they saw a tornado touch down. We should take it with a grain of salt." Sid responds, "It makes sense. Eyeballs work. It all fits in." This report from the public fits with what they expect, so this report is given the credibility that it might otherwise lack. (Field notes)

As this episode indicates, verification is not self-evident, but must be processed organizationally to fit into the logic of accountability. One forecaster put the matter directly:

> Now there's so much pressure on proving the numbers. I think we're way too liberal in how we interpret things. Someone [a member of the public] says 50–60 mile an hour winds. If we didn't have a warning, we'll call it 50. If we do, we'll call it 60. Because of course you want to verify. You want to be right. If you put out a warning for Mohawk County, the next day you'll spend two hours making fifty phone calls trying to find something to verify the warning. If you didn't put out a warning, are you going to spend three hours making fifty phone calls? No. (Interview)

As is often the case in organizational culture, strain is expressed through joking.[29] Forecasters kid about making the world fit their forecast by whatever means necessary:

> When we need some damage, I'll take off a few shingles. I'll go out and verify whatever you want. (Field notes)
>
> : : :
>
> During severe weather, we send [an intern] out with a chain saw [to cut down a tree for verification]. . . . After a big snow, we send interns out to measure snow, and if they don't come back with the amounts we forecast, we send them out again. (Field notes)

Such joking is accepted because they believe that as honest and competent professionals serving the public is more important than verifying statistics. They keep doing their job as best they can and then fudge the numbers if necessary.

The Politics of Political Boundaries. If there is no outright cheating, frequently a proactive search for supportive information occurs. This search is built into the procedures, and, given that local offices issue warnings by counties, which are political units, verification also is judged by county. A tornado that touches down just outside of a county boundary does not verify for the county that it barely missed. Being on the wrong side of the road, it constitutes an organizational failure. Given that counties have different resources and different attitudes to communicating with the NWS, responses to warnings vary, leading to forecasters not warning some counties of the possibility of minor severe weather:

> Sean reports at the staff meeting: "The day shift has to make the callbacks the next day to find out if the events are verified. It has to be done the next day. It needs to be done every time there are unverified warnings. You have to be specific, Do you have reports of structural damage? Do you have any reports of trees down? Do you have any power lines down? The more specific you are the more likely you will get a response. One needs only one form of verification. If you've got a severe weather warning for Orange County, [the head of Emergency Services] has computers all over the county." Sean emphasizes that Orange County is likely to verify severe weather, but other counties are not, based on the structure of emergency services. One suburban county attempted to integrate fifteen counties into a centralized warning system, but most counties in the less-populated areas have since dropped out. Sean notes, "McKinley County, you're not going to get anything." Sean notes that even if they warn for a part of a county (e.g., eastern Orange County) and there is a confirmed tornado in another part of the county (western Orange County) that verifies the warning. He suggests that by issuing warnings for longer periods, they might increase their verification scores. He concludes, "I realize that verification is not supposed to be in the front of our heads, but that is what Congress looks at." (Field notes)

The location of a storm—and the organizational commitments that are tied to that location—contribute to the likelihood that events will become verified. Key to knowing are those practices that are linked to the politics of place.

Proactive Work. As Sean's quotation emphasizes, forecasters are proactive, but they are proactive in verifying their warnings, not for the areas for which they didn't warn. A forecaster at the SPC explained, "When the events are marginal, [local offices] have a certain influence over them. If they have issued a warning, they will ask around. If it is an event they haven't warned for, they may not ask as hard" (Field notes). One forecaster, commenting on the relatively low severe weather verification scores of his office, noted, "We definitely need to be making more calls. We definitely need to be more proactive in finding out what is going on. Sometimes it's just the nature of the game" (Interview). Another added, "That's something that other offices do to pump up their verification" (Interview). Consider these cases:

> Ritchie, the college student intern, calls local police departments and emergency offices in areas for which warnings were issued. Without him these calls would not have been made. As a result, almost all the warnings were verified.
>
> Sid remarks: "I think we've hit every county we've had a warning for. . . . I think we're going to bust in Dilling and Smythe." However, Ritchie says that he will keep calling Smythe until they admit they have a tree down. Eventually he gets verification from both counties.
>
> Sid says, "So far I'm batting a thousand."
>
> Ritchie tells him, "You missed Blaine County. 990."
>
> Sid responds, "We might get a report tomorrow. We can scrounge up something. We can find something."
>
> Ritchie jokes, "No one should bat a thousand. . . . Trees and power lines down. I think that is a generic thing they are telling us. It counts as verification."
>
> Sid laughs, "Blaine, they'll find something."
>
> Ritchie tells me, "He'll go out with his car and hit a tree." (Field notes)
>
> : : :
>
> After issuing a warning for southern Foster County, Bert calls the local police and asks if damaging winds occurred. Ritchie then calls a nearby reporting station and reports that in a previous

storm cell (that wasn't warned for) there was a tree down and one-inch hail. Ritchie suggests, "We could fudge the time."

Bert responds humorously, "That's a big fudge."

Sean tells Ritchie with certainty, "That's a bust. We didn't get a warning out."

Bert jokes about southern Foster, "Anybody else who might have that information. Maybe Miss Cleo?" (Field notes)

: : :

Officials from Blaine County call to announce that trees have been reported down. Randy jokes, "We didn't warn for that. Disregard. They were dying anyhow."

Ritchie reports a call from Greg's wife in McMartin County, "We don't want to hear from her. Tell her to say that it was 57 miles per hour until the warning goes out, then she can call back and say, 'No, it was 58.'" [Severe thunderstorms verify with 58 mph winds]. Ritchie tells Randy, "I didn't call Foster County. I don't want to hear from Foster" [which they hadn't warned for].

Randy comments that they received a report of 49.3 mph winds over Lake Michigan which requires 50 mph. Byron jokes, "Could you round that up? That's God playing around with our heads." (Field notes)

The strategy involves both investigating and not investigating, hoping for confirmation. Although contact is often after the fact, offices sometimes call counties before severe weather warnings are issued to prime them for impending bad weather. Like lawyers with guilty clients, forecasters learn to ask the proper leading questions, "Would you say that the hail was quarter size? [not dime size]" or "Would you say that the tornado was spotted five to ten [not fifteen] minutes ago?" The naively "wrong" answer can create organizational distress.

Although verification might appear a simple matter, the absence of material traces of short-lived weather events, coupled with the inability to verify these events through the same technology by which they were originally noticed, challenges institutional practice. That organizations are committed to accountability makes decisions social and political as well as scientific.

The Politics of Verification. Although some peculiarities are associated with operational meteorology because of its role as public science and because of the immediacy and human-level quality of its predictions, issues of verification are found in many occupational worlds that make explicit or implicit claims about the effects of natural processes or human interventions. If weather has more salient issues because of its routinized and everyday qualities, it is not alone. We search for ground truth on which we can judge what has occurred and then create procedures of verification by which we can share those assessments with others in ways that reveal our credibility and demonstrate our accountability. Our claims to competency must be verified, establishing the legitimacy of expertise.[30] When, as in the case of the weather service, rewards are given to offices based on their "success," strategies develop to insure that this success rate is as high as possible. From the standpoint of the larger organization, this is not necessarily a bad outcome, as they, too, are judged on the success of the claims of their local units.

Procedures must be developed to determine the adequacy of these claims, and then strategies are developed to "work" these procedures for best advantage, providing favorable assessments of practitioners. How close is close? What level of confidence, statistical or otherwise, is sufficient for determining that one's prediction is correct? Each occupation wishes to determine its own rules of practice. Meteorological verification needs to be compared and contrasted to other worlds: tests of environmental pollution, petrology, climate modeling, or earthquake measurements, not to mention education, economics, and medicine. Each has institutional demands placed on the production of predictive claims and strategies for measurement of those claims.

Verification is an organizational practice. Although it is linked to the "real world," it is not a transparent window into that world. The adjudication of truth is a collective practice on which confidence depends. Those who claim a vision of the future view their creations through a window whose assumed transparency directs attention away from its distortions and shadows.

The F5 at La Plata

In late April 2002 a deadly tornado devastated much of the Washington suburb of La Plata, Maryland. This was a powerful tornado, but how powerful? How will this rare East Coast tornado enter meteorological history?

Every year approximately 1,200 tornadoes touch down in the continental United States. These tornadoes cause more that $400 million in damages and some 75 deaths. When a tornado hits, the local weather service office must determine its intensity. Tornadoes are rated in their destructive power on a six-point scale, the Fujita scale, named after the late University of Chicago meteorologist Ted Fujita. His scale ranges from F0 to F5. An F0 is a "gale tornado" with winds between 40 and 72 mph (below hurricane-strength winds), causing light damage. An F3 tornado is "Severe" (158–206 mph). An F4 is described as "Devastating" (207–260 mph) and an F5 tornado is an "Incredible tornado" (261–318 mph), with the added explanation, "Strong frame houses lifted off foundations and carried considerable distances to disintegrate; automobile-sized missiles fly through the air in excess of 100 yards; trees debarked; steel-reinforced structures badly damaged; incredible phenomena will occur."[31] Despite public beliefs and the reference to wind velocity, the Fujita scale does not measure tornado strength as such but only damage. Given that tornadoes with their evanescent quality cannot be directly measured, the assumption is that the amount of damage is directly correlated with tornadic strength, and so each F-level has an estimated wind speed. The extent of debris, however, depends on the location of the storm. Storms that pass through fields are likely to be judged differently from storms that hit towns. For smaller tornadoes, F2 or below, the local office makes the assessment. Ratings of these windy storms are of little political or scientific interest. For more intense storms, while the local office makes a first assessment, the NWS sends a Service Assessment Team that reviews the storm damage and publishes a report that documents the storm.

On April 28, 2002, a series of strong storms spread across the Midwest, South, and Mid-Atlantic states—a major outbreak of tornadoes and severe thunderstorms. Local offices issued 393 severe thunderstorm warnings, ninety tornado warnings, and the SPC issued fifteen watches that Sunday. The most intense tornado carved a 64-mile swath across southeast Maryland, producing extensive damage in the suburban Washington community of La Plata. Three deaths and 122 injuries resulted directly from the Maryland storm. Property damage exceeded $100 million. Kevin Grinder, a dispatcher with the Charles County emergency services reported:

> The first we heard of the tornado was from the crew of Ambulance 29 from Hughesville. . . . A few seconds later, we felt it. I've never seen a tornado before, and I hope not to again. . . . The high winds pulled open the doors of the building. We have a drop ceiling, and the pressure pushed the ceiling tiles up about 6 inches. I've heard people say a tornado sounds like a freight

> train. I don't know whether that's what it sounds like, but we could feel the whole building shake.[32]

Shortly after the tornado hit, the local Weather Forecast Office in Sterling, Virginia, rated the storm as an F4, but an hour later raised the assessment to an F5, based on building damage. This dramatic report was disseminated to the public. The La Plata tornado was one of those rare tornadoes (on average one each year) that receives an F5 rating. Given its destructive force and its proximity to Washington, the tornado received extensive media attention. Eventually when the national office made their assessment the storm was downgraded to an F4.

The devastation and death were painfully real, evident to any observer, but as a scientific matter, what *was* the tornado? More precisely, how did the meteorological community respond to the assessment.[33] Changing a tornado's rating is rare, although not unheard of.[34] My concern is not to determine the proper rating, but what the incident suggests about the politics of verification.

During the storm and in its aftermath I was observing at the Storm Prediction Center in Norman, Oklahoma. I watched the forecasters issue a watch for the Mid-Atlantic states, and followed the radar images as the storm traveled through Virginia and Maryland. As I noted, forecasters cannot determine whether a tornado touched down from the radar images or, if it did, the extent of damage. But these men were tornado connoisseurs. Many were storm chasers and had seen numerous devastating storms on radar, in the field, and through their aftermath.

Within a day, shoptalk spread in Norman—and throughout the NWS—that the local office erred in their hasty evaluation of the La Plata tornado as an F5. Meteorologists, viewing images of destruction in La Plata, concluded that it didn't compare with the nation's most powerful tornadoes, notably the 1999 Oklahoma City tornado, the 1990 Plainfield tornado, or the string of seven F5 tornadoes that hit Alabama and Ohio on April 3, 1974. After an F5 tornado, the site should have been leveled, but that was not the case in La Plata. Leaves remained on the trees, mailboxes were standing, and play sets were not crumpled. Evaluating tornado damage is not based solely on an assessment of the destruction but an assessment compared with the collective memory of similar events. Meteorologists are like doctors, stock market analysts, or detectives who compare the cases they are confronting with memorable cases of times past.

Some forecasters believed that the tornado was rated an F5 because it hit near Washington, DC, and the evaluation was a political decision, "pressured from media hype" (Field notes). Local residents often hope to be part of meteorological history, and an F5 tornado would make their experiences historic.[35] A common view was that the Sterling office "jumped the gun, and the media picked it up," concluding that "we do not work in a vacuum" (Field notes). As sociologist W. I. Thomas[36] understood in recognizing the importance of the definition of the situation, the F5 rating became real because it was published. A related view suggested that the staff at the Sterling, Virginia, office had little experience in dealing with tornadoes of this magnitude, and, while it was the worst that they could imagine, it did not compare with the most powerful storms in the Plains. A certain regional pride was evident—possibly in the East, but certainly in Tornado Alley in Oklahoma. One staff member suggested that "the consensus around here is that this was not really an F5. I'll call it the East Coast bias" (Field notes).

While one forecaster explained, "I'll keep my mouth shut," sensing an air of politics, others were voluble. A scientist at the National Severe Storm Laboratory stopped me two days after the tornado, describing the controversy and informing me that "the war may already be lost," meaning that he felt that the La Plata tornado would remain an F5. The decision, while ostensibly scientific, might influence the amount of federal aid available to the Maryland suburb. He explains that "Some people are upset because it distorts the database. It's the worst thing [Sterling forecasters] have ever seen. There's the hype aspect. I know there is a lot of pressure. I think they are in a difficult situation." One forecaster, assessing images of La Plata, commented, "It doesn't pass the sniff test with a tornado going through a town of 3,000" (Field notes). As the discussion continued, forecasters emphasized that East Coast tornadoes would be rated lower if they hit Tornado Alley and that, despite formal assessments, "There is no uniformity" or "It's pathetic how tornadoes are evaluated."

Meteorologists felt that judgments of tornadoes are as political as scientific, and they note that forecasters receive no formal training in making damage assessments. The period of 1970 to 1986 was labeled "the glory days of wind damage evaluation"; during that period the assessment of tornado damage was funded by the Nuclear Regulatory Commission, which was worried about the effects of tornadoes on nuclear power plants. That funding has expired, and today "everyone and their brother goes out and rates it according to how they feel" (Field notes).

Debates continued throughout the week, as did media attention. At one point I asked if there had been any F3 tornadoes on April 28; one forecaster chuckled, adding, "the one in Maryland." Five days later, on May 2, severe weather again threatened the Washington area, and the memory of La Plata—F4 or F5—was so raw that schools closed early; a forecaster sniffed, "they are just going crazy." Another forecaster worried about issuing a watch: "I wonder what will happen if I issue a box. Will they go sit in their basements? There'll be an evacuation." A third meteorologist notes that the media are hyping the threat, "They should just head to the mountains and jump in a hole." Yet another remarks, "Terry [the lead forecaster] has evacuated half of the East Cost." Guy adds sarcastically, "And they say no one reads our products. That's power!" (Field notes). Fortunately on May 2, no tornadoes hit Maryland, but the community was primed.

Finally on May 7, the special assessment team, relying on a prominent wind engineer to examine structural damage, announced that the La Plata tornado was "officially" an F4—a devastating tornado with wind speeds of up to 260 miles per hour. However, given the earlier assessment, the NWS team was careful in their wording, attempting to salve the community's pride,[37] writing, "'Regardless of the final "F-strength," this tornado outbreak was deadly, destroyed property and disrupted many lives throughout southern Maryland,' said team leader John Ogren. 'Tornadoes can kill at any strength, and can strike anywhere.'"[38] Ogren told me, however it might have been received in Norman, "It was not a popular decision with the community," who had already printed rhyming t-shirts reading, "Community Spirit is Still Alive After an F5." For the residents of this small town, surviving an F5 was a badge of honor.

Ogren informed me that the team felt that it was actually only marginally an F4, and could legitimately have been rated as either F3 or F4. He noted, "if they [the Sterling, Virginia office] had rated it an F3, we would have kept it there." They went with the F4 designation to support their colleagues and perhaps to assuage the community.

So, after internal debate and bureaucratic action, the La Plata tornado was cemented in history as a major storm, but not among the elite "Tornadoes of the Century." The clash of social and scientific was enshrined in the debate over just how powerful this freak of nature was.

Social Weather

The constructionist perspective is well established in the social sciences.[39] This approach in its most powerful form suggests that the meaning of the world is not to be discovered but to be imagined, a function of interests of practitioners, coupled with their resources. Science workplaces become sites of knowledge production[40]—creation, not description. This view, the Strong Programme in science studies,[41] argues that science must not be privileged, but should be treated as any other area of human endeavor, taking an agnostic stance to the truth of scientific claims. These claims are political by being connected to domains of power. Of course, one should not overemphasize these political pressures; they are not the sole criterion through which knowledge is created, and typically not the most significant, but science and technology are shaped by politics, influencing how weather is seen and reported.

While weather consumers assume that operational meteorology has objective standards, separate from the social needs of scientists, supervisors, sponsors, and clients, such a sharp division is not fully tenable. The view that meteorological choices are simply a result of impartial knowledge and neutral labor does not hold up under scrutiny. As Michel Callon writes of engineers, "Whether they want to or not, they are transformed into sociologists."[42] Rik Scarce explains, "As social actors, [scientists] are embedded in a social structure, and to be effective in their discipline they must familiarize themselves with the social forces all around them."[43] Forecasters operate in a social world, as well as a physical world. They translate that physical world, as they understand it, into a social world.

I begin with their joking culture. If a forecaster writes the forecast, perhaps he can create the weather. Of course, no one believes this, but it is *close enough* to being true that it sparks flippant remarks:

> It is a cold April day (37 degrees), and Joan, the administrative assistant, says to Don, "You guys have to do something about the weather." Don replies, "I wanted it to be seventy degrees." Joan retorts: "Well, make it happen." (They laugh) (Field notes)
>
> : : :
>
> The day of the MIC's wedding has beautiful weather. Randy jokes, "We got the weather through for you." Sid replies, "Do we get a raise for that?" (Field notes)
>
> : : :
>
> Sid jokes early in my research: "You want to know our secret. It's 1-800-CALL-GOD." (He laughs) (Field notes)

More seriously forecasters see themselves as responsive to public expectations. This may result in forecasts that tip either toward the optimistic or the pessimistic:

> George explains that staff members have different preferences in forecasting. One difference is whether they are optimistic or pessimistic. His choice is that you "don't say partly cloudy, say partly sunny." Although they have identical meanings meteorologically, he explains that people hear the first and think there is a 40 percent chance of rain, but if they hear the second, they think there is only a 15 percent chance of rain. (Field notes)
>
> : : :
>
> Gavin tells Cal, "My gut is to go pessimistic until tomorrow." In other words, he doesn't want to be blamed for saying that it will be nice, when it may remain drizzly and overcast. He tells me, "There is a politics of a busted forecast." Cal adds, "If it's better than what you forecast, they're not so upset. Sometimes that enters into your thinking. If you're pessimistic, people won't care." (Field notes)

Whether one selects to be pessimistic or optimistic, both involve adjusting the forecast to the imagined desires of one's audience. Perhaps both optimistic and pessimistic forecasts are legitimate, but they affect people differently, both at the time and subsequently if they are proven to be incorrect. The pessimist protects himself from failure, while the optimist wishes to lighten the public's emotions as they plan their day.

Choices are made in light of public expectations for holidays and weekends:

> Byron tells me that there are more blizzards forecast the week after Christmas than the week before. He tells me that forecasters don't want to harm people's holidays [I was also told that it was because merchants are angered by mistaken snow forecasts the week before Christmas]: "You need to be absolutely right. . . . I think there are times that [forecasters] will not call something [i.e., bad weather] because of weekends, holidays. What is being put out is not being put out by robots. We are sensitive to what is going on. There are mitigating circumstances, such as holidays or the Taste of Chicago unless your hand is forced. . . . Every

> business has its backroom. Its lingo, its secrets, and then there is what the public believes." (Field notes)
>
> : : :
>
> Randy explains about forecasting showers for Memorial Day: "Not exactly the kind of forecast you want to put out going into the Memorial Day Weekend." However, given the weather guidance, he feels that he has no choice. (Field notes)

In many cases, what is good weather for some is harmful for others[44] farmers may need rain, while contractors hope for sun—but on holidays and weekends a consensus is clear.

In response to a local media claim that some winter forecasts are "political," designed to inflate the amount of snow to increase federal funds if the city is designated a disaster area, Chicago forecasters emphatically deny that economic or political pressure influences their calculations. Because of their insulation, the absence of any direct manipulation is surely correct. Still, forecasters are well aware that their choices—winter storm watches, wind chill, heat advisories, aviation visibility—have both economic and social effects. A phrase that I heard on several occasions, "putting out the forecast of least regret," refers both to the forecast by which the public will be protected (evacuating towns early from a possible hurricane threat) and those that protect the forecaster. These forecasts are sometimes labeled "political" in that meteorologists attempt to manipulate the public to act in ways that they think necessary, even if the forecast is slightly stretched. In contrast to verification statistics that ignore the social consequences of forecasts, meteorologists, taking social life into account, may "overwarn" the public. They justify their slanted forecasts because "People could lose their lives. What if we're wrong? Your name is on the forecast" (Field notes). Another noted, "if it's rush hour, there's going to be a lot of traffic over a major holiday weekend and the greater potential for problems, then I would forecast a warning" (Interview). A group of forecasters in my presence discussed manipulating hurricane paths, not disseminating the most likely meteorological prediction, but motivating citizens to follow instructions. This apparently happened with Hurricane Floyd in September 1999, producing the largest peacetime evacuation in American history, removing much of the coastal population of Florida and Georgia. If an optimistic forecast were wrong, corrections would be too late, and so fudging was legitimate. Since hurricanes are the primary weather event that requires evacuation (some river floods do as well), it is here that social control in forecasting is most evident. Of course, a manipulation that was too obvious could provoke a scandal, so the steering must be subtle but sufficient to provide for public safety. As one forecaster revealed, "They will fudge on the side of evacuating." Another comments, "You may shade your forecast because you want a particular response" (Field notes). For their part, the SPC sometimes includes a threatened metropolitan area in a watch box earlier than they might otherwise do to prime citizens of danger, what they term "playing the population game." As one forecaster remarked in his conference call to the St. Louis office, then at the edge of a threat, "I would like to include St. Louis City in [the watch] to enhance awareness as much as I can" (Field notes). Population affects weather forecasts, whether or not it should in purely scientific terms, reminding us that operational meteorology is a *social* science:

> Terry issues a severe thunderstorm watch box for southwest Texas, but doesn't think it is very significant because the area has few residents. He focuses more on potential thunderstorms in New York City. He tells me, "*In an ideal world*, population wouldn't weigh into weather, but we're not in an ideal world. Those things have to sway in a little bit." He explains that regions have different criteria for severe weather. Small hail is taken for granted in the rural Midwest, but would be worth a watch in New York City. He adds, "You put the exact same storm over Western Pennsylvania that you put over Central Oklahoma, and you get a very different reaction from the public." (Field notes)

In avoiding the worst-case scenario, one covers oneself, and as a consequence more warnings may be disseminated that would be necessary in terms of standard threat criteria. In a purely scientific frame (Terry's "ideal world"), these social factors would not be relevant, but meteorologists consider how publics will respond to their forecasts.

Along with the desire to have the forecast produce the desired response, a forecast aesthetic exists. Here I am not referring to the writerly aesthetic, discussed in chapter 4, but to how the meteorologist should construct the forecast for the public. I was sitting with a forecaster who faced a weekly forecast in which every day was forecast to be "partly cloudy." He decided to change several of the days in the extended forecast to mostly cloudy and mostly sunny to give the forecast more diversity, and to make it clear that he had been doing his job. Another forecaster told me that he never writes that the likelihood of rain is 50 percent, selecting either 40 percent or 60 percent, so that readers won't think that he can't decide. In fact, when forecasters, at least in the Midwest, have minimal confidence as to whether it will rain they claim a 30 percent chance of rain, since measurable rain occurs on three of ten days in Chicago.

One forecaster decided to forgo rain in his forecast, but he told his replacement, "There is a secondary front back here that you might keep an eye on. If it holds together or strengthens, you might put in sprinkles. We're going to have enough rain in the forecast for the next few days" (Field notes). He didn't want to depress his public. On another occasion, a forecaster sardonically commented about rain showers that he decided would end at the state border, not affecting his state or his forecast responsibility, "Sometimes it's interesting that weather follows the state line" (Field notes). An implicit sociology of weather forecasting recognizes the need to play to one's audience.

Snow Job. I had no idea about the challenges of measuring snowfall. After each snowstorm, the public is told that a number of inches of snow fell. But what does this mean? The measure depends on organizational choices. First, one must determine when a snow event begins and ends. This is not a simple matter as snow often starts and stops and falls in flurries. What are the temporal boundaries of an event? Second, one must select the location at which to measure. Too close to a building affects the amount of snow, similarly a snow that is driven by a strong wind may fly horizontally and a measuring instrument with a small opening may miss some snow. Third, is one measuring *snow amounts* or water content? A wet snow will be deeper than a dry snow, even if the number of snowflakes may be comparable. Fourth, one must assume that the measurements are accurate. Sometimes they are not available, forcing forecasters to estimate. Finally, how often should one measure snow? Does one wait for the snow to end or measure every fifteen minutes, summing the amounts. Snow compacts and melts. For each of these choices organizational rules have been established, changed and shaped by local practices.

One day when I arrived I found the Chicago office in turmoil. A heavy snowfall hit the city and the observer at O'Hare Airport took the hourly totals and added them together. He provided the Chicago office with this amount, and they distributed a public information statement announcing a 17.7-inch snowfall at O'Hare, a very heavy amount. The math was correct, but the procedure was not. The snowfall amount is based on summing *six*-hour totals,[45] which would have put the snowfall amount at 12.0 inches, because of the melting and compacting of snow. The office received media calls all day. Said one forecaster, "All Hell broke loose. Once it was sent out, it was awfully ugly. I don't think I've ever seen the phone ring so much. Whenever the media senses something they can jump on." Finally the MIC distributed a statement explaining the mistake. One meteorologist groused, "It made us seem that we didn't know what we were doing." They made an organizational error, but the public didn't know why the figure was wrong, because the idea of a snowfall total seemed to be objective and obvious. As one forecaster commented about measuring precipitation, "as you know more things, the simple things become more complex" (Field notes).[46] Even the simplest measurement depends on organizational rules.

The Politics of Forecasting. The idea of a political forecast was a particularly sensitive subject in Chicago, a city in which almost everything is political. The archetypal political forecast, however, was that of Sean Boyd, a radio weather forecaster fired for predicting rain for a Rush Limbaugh event. The station management, sponsoring the event, wished him to fudge the forecast, predicting the possibility of sunshine, rather than a chance of rain, fearing that the forecast might keep listeners at home. When Boyd refused, he was terminated. On the day of the picnic, "the heavens opened and it poured."[47] There is an obdurate reality after all, at least for conservatives.

Forecasts, like maps,[48] are political because they accept political boundaries as their primary reality. They link weather to the state, nationalizing it.[49] There is Swedish weather, Florida weather, Chicago weather, and so forth, and these are reported as such. Meteorological organizations are national, supported by the state, and local offices in the United States have responsibilities for a set of counties.[50]

While there is "always politics in everything," forecasters are shielded from overt political pressure. Still, some subtle reminders remain. Local offices are judged on how well they forecast for the largest cities in their County Warning Area. Authorities pay relatively little attention to what happens in the boondocks:

> Arthur informs me that where storms hit matters in terms of how the forecast is viewed, and this affects their care in forecasting: "If we predict severe weather, and the storms miss Prairie City, it looks like a big busted forecast. The only exception is that if it misses the city and annihilates some small town and kills

> people, and then people pay attention. Otherwise it's seen as a wrong forecast." He tells me that there might be three small thunderstorms that hit Prairie City with 3 inches of rain and that is a major event, even though in meteorological terms the damage is minor. He jokes, "I say if a tree falls in the forest, it doesn't matter, but if it falls outside of Channel 13. . . . " (Field notes)

This attitude helps to explain why forecasters are relatively unconcerned about weather in rural areas, less media-saturated, and where errors cost less materially and socially. Perhaps these rural citizens are more self-reliant given their folk knowledge, perhaps there are fewer of them, or perhaps they are simply less politically potent. The Chicago office occasionally receives complaints from officials in their western counties, but those complaints do little to change the forecasting priorities of the office. As one meteorologist who was raised in that area noted, "It seems as though the forecasters just concentrate on Cook County, and not the rest of the area. Henry Potts, the Bridgestone County Director, he says that [severe weather] hits us, and it's far out east before we see anything on the screen. . . . Because I'm from the area, I guess I take it too personally that they don't give service" (Interview). The office does not have many observers in this region, and so the area does not have high psychological salience. Lacking data, forecasters tend to underforecast. As one forecaster noted, "Having an observation point makes it real psychological. It makes it real. [If there is no observation point], so what if you bust, who's going to know" (Field notes). In contrast, the office discusses heat advisories with city officials in Chicago, and developed criteria for warning after the deadly heat wave of 1995.[51] Other political jurisdictions do not have the same influence in the creation of forecasts, nor, of course, do they have three million citizens potentially at risk.

The Storm Prediction Center with its national mandate realizes that some areas are sensitive to watches. Sometimes this is a function of recent tragedies that prime an area for severe weather, but other times it is a consequence of politics. I was told that the Huntsville, Alabama, office was almost closed during the agency reorganization, but because of congressional pressure, the office remained open. An SPC forecaster told me that as a result, he decides whether to issue a watch for the area with rare care. He explains, "I hate to play politics. . . . The media, the public in Huntsville are very aware of weather, and so they get very good service, even from us." Forecasters describe northern Alabama as a "politically sensitive zone." Huntsville is not alone in this. Evansville,

Indiana, which lost its office to Paducah, Kentucky, receives good service because of the closure of their office. "People in Evansville didn't like that, and five years ago the Paducah office didn't handle a tornado in Evansville very well, and there was all Hell to pay" (Field notes). Since that time, to pacify the good people of Evansville, a new local radar was installed, feeding information directly to the Paducah office. Political process and social pressure has its effects, particularly with an organization, such as a government bureaucracy, that is embedded within a system of power.

Record-Worthy Days. One of the arenas of fascination for public and meteorologist alike is that of records. Like living through severe weather, records provide people with a set of temporal benchmarks.[52] It inserts them into history, if only momentarily. They were where the action is. The bimonthly magazine *Weatherwise: The Magazine about the Weather* documents weather records, some rather esoteric:

> [March 2003] Portland, Oregon, noted measurable rainfall on 27 days during the month, breaking its March 1961 record of 25 days.[53]
>
> : : :
>
> [April 2003] With a high of 92 degrees F on the 13th, Bismarck, North Dakota, marked its earliest 90-degree temperature on record.[54]
>
> : : :
>
> [April 2003] Kahului, Hawaii, received a record-low April rainfall, totaling 0.01 inch (the norm is 1.75 inches) and breaking its April 1990 record of 0.06 inches.[55]

Admittedly this is a journal aimed at weather enthusiasts, but that these measures are documented reminds us of the use of extremes in constructing memory, a phenomenon also found in sports talk and politics.

But what do records mean? Records presume that we can select categories of measurements that belong together. We keep records of high and low temperatures or the amount of precipitation for each day of the year. The media (and the weather service) may report that this is a record high or low temperature for the day (say, March 3), although they are not likely to report that there was a record rain, record snow, record windchill for that particular day. Those records are monitored on a monthly, seasonal,[56] or annual basis. Temperatures are monitored on a daily, monthly, seasonal, and annual basis, but not on a weekly basis. How we view time determines what we recall.[57]

In addition, the location of the measurement must be chosen, even if it is not reported. Currently in Chicago weather is measured at O'Hare International Airport, but the official location has moved several times over 130 years, making historical comparisons problematic, as some locations are closer to the weather patterns of Lake Michigan, and, of course, instruments evolve as well. For instance, August 2001 was declared to be the second wettest August on record (a total of 12.25 inches), but it was only so because O'Hare was drenched with two heavy rainfalls, one of them over 4 inches. These rains were localized. If the measurement had been taken at Midway Airport the month would not have been meteorologically notable (the rainfall total at Midway was under 7 inches). The rainfall total in the south Chicago suburb of Lansing was only 3.5 inches. One forecaster told me that on occasion the official readings of snowfall change from O'Hare to Midway as "the political powers that be swap back and forth, depending on how they did [on snow removal]" (Field notes). If snow removal went well, they would select the heavier snowfall amount; if it didn't go as well, they would select the smaller amount to indicate that the snowfall was not a major event.

When is a record a record?[58] Who decides?[59] The answer is when an organization claims that it is. These issues are typically invisible because of the assumption that records have an obdurate quality, but under some circumstances they provoke notice. The most dramatic example occurred when a media forecaster reported that Chicago had a record cold month from mid-May to mid-June, causing other media outlets to call the local office. But while the data are available, the time from mid-month to mid-month is not a "real period." Monthly records should begin on the first of the month. One forecaster described the record as bogus, even though the numbers were correct. A forecaster on duty that afternoon, said that "After the [newspaper] went with it, we got calls all day." This forecaster pointed out that because of the heat in the first week of the month, May was actually slightly warmer than normal! Was this a record? It was because a newspaper printed it, and the surprising cold encouraged someone to examine the statistics, but it is unlikely that such a record will be checked unless the weather seems *intuitively* unusual, that is, record-worthy. Such cognitive focus occurs elsewhere. When we believe that something unusual is afoot—stock market gyrations or sports streaks, for instance—we compare them to historical records, reminding ourselves that we live in unusual and memorable times.

Ground Truth

If air is a metaphor for the clear and transparent, the practice of meteorology demonstrates the limits of the analogy. Yet, meteorology claims that prediction is possible. Central to the work of the meteorologist is the ability to assert competence and trust through a structure by which these claims are defended. This is the process of verification—the organizational contention that forecasters got it right. But what is *right*? The process of verification is not as obvious as the term suggests. Temperature is a function of where and how it was measured. Snowfall amount depends upon multiple choices, erased when the public is told that 8 inches of snow fell. In effect the event itself is created after the wind dies down. Of course, meteorology is not alone in its problems with assessment. Sports statisticians must determine the length of a home run and news media (and police departments) must determine the size of crowds of demonstrators, both of which require procedures to generate the proper statistic.

This process is evident in rating tornadoes. We have a strong desire to categorize those events to which we are exposed, making them memorable, and allowing us to create collective memory. We do this with hurricanes, snowstorms, and with tornadoes, as with sports, politics, economics, and entertainment. Each domain, meteorological and otherwise, has its own challenges, but tornadoes because of their brevity are particularly problematic. The Fujita scale that is currently accepted attempts to compare tornadoes. The scale could be ordered in any way, but it is structured to establish a few memorable events at the tail of the distribution of cyclonic winds. If most tornadoes were rated as F5, rather than F0, such events would have less significance as memory hooks. The fact that on average each year produces one F5 tornado creates a natural celebrity, ready to be memorialized on the Weather Channel's "Storm Stories," the *People* magazine of the air. The number of hurricanes of truly historic—memorable—proportions is equally low. We recall them by name, forgetting their weaker colleagues, whose names are used again and again, reincarnated until they too produce havoc and memory.

6 A Public Science

The astronomer is, in some measure, independent of his fellow astronomers; he can wait in his observatory till the star he wishes to observe comes to his meridian; but the meteorologist has his observations bounded by a very limited horizon, and can do little without the aid of numerous observers furnishing him contemporaneous observations over a wide-extended area.

James Espy, 1851[1]

: : :

As James Espy notes, meteorology is an early example of big science, embodying the collective production of knowledge that became well established during the twentieth century.[2] Meteorology is collaborative, both in its extensive data gathering and in its need to communicate to a public outside the narrow confines of its professional community. It requires forecasters to be translators to make their local expertise available for larger audiences.[3] Perhaps it is this feature, its wide, curious, and insistent public, that has contributed to the decline of meteorology as an elite science. The absence of meteorology departments in most elite private universities is striking. The field may be too close to popular entertainment for scientific comfort. Likewise, the direct government involvement—the extensive government job market, a market for undergraduates—places meteorology slightly outside "real" sciences. With its public service aspects, it is closer to the more applied fields of engineering, horticulture, pharmacy, nutrition, and agricultural sciences. The government encouragement to state universities to establish meteorology departments further stigmatized the field as *a domain of applied practice*.[4] In the current age, "real science" requires the credential of a Ph.D.

In this chapter I focus on the organizational features of meteorology, particularly the position of operational meteorology as government-sponsored activity. The organization of operational meteorology highlights the tension between autonomy and control. As Leslie Sklair argues,[5] scientific domains are integrated into society and cannot be separated from the organizational infrastructure that they help create. Not only do sciences have managers, but they also have audiences. The presence of an audience produces demands for expertise. I examine the implications of the organization of the National Weather Service as a governmental agency devoted to public science. I then explore how this occupation is linked to a set of clients on the output boundary[6] of government meteorology—the public, the media, and private firms.

Close Enough for Government Work

However we might judge the character of individual offices, they are each part of a large government bureaucracy. The National Weather Service is a branch of NOAA, the National Oceanic and Atmospheric Administration, part of the Department of Commerce. The agency, headquartered in Silver Spring, Maryland, employs approximately 4800 workers. If these employees are scientists, they are also bureaucrats, not that most would admit it. They labor in a vast agency, even if, as a result of civil service policies, their employment is more secure than it might be in the private sector.

The National Weather Service, like each local office, has an organizational culture, both official and unofficial. I have quoted the first sentences of the mission statement, but the statement concludes, "NWS provides data and products for a national information database which can be used by other government agencies, the private sector, and the global community." The mission statement connects the weather service with other institutional arenas, notably the economic sphere, perhaps not surprising given its current placement in the Department of Commerce. Weather information is linked to the needs of corporate capitalism in protecting property, enhancing the national economy, and providing data for the private sector. Some forecasters felt that the movement from writing forecasts to manipulating databases was part of a process by which the agency served corporate clients, backing away from contact with citizens. The rhetoric is moving toward emphasizing corporate efficiency. One administrator explained that the goal of the agency for the next year is to work with the airline industry to achieve a 10 percent reduction of late arrivals and departures through better weather information. He explains, "The higher echelon are really looking closely at bang for the buck. Not just spending on new toys, but a return on investment" (Field notes). Getting the forecast right is treated as child's play. Government is judged like a corporation, and in the process some feel that "the human factor is fifth or sixth" (Field notes). The weather service has transferred some of its functions to private companies, such as agricultural forecasting, and there have been some suggestions that the weather service not place forecasts on their website, so as not to compete with private companies.[7] Some companies feel that, with the exception of warnings, the National Weather Service should do no more than provide raw data to those who contract for that data. Were it not for the expense of maintaining the meteorological infrastructure and the legal liability of severe weather warnings, the agency itself might have been privatized, and attempts have been made over the years.[8]

In addition to being an agency that operates as an adjunct to industry, the NWS is also a quasi-military organization. Both the linkage and the tension between science and the military have been well documented. Military needs shape scientific work, times of conflict are often productive moments for science, and physical science budgets are tightly controlled through government largess,[9] creating knowledge and manpower to use for other projects.[10] Strain exists about whether the model of the NWS should be the academy or the military; elements of each remain. But the fact that the director of the agency during this research was General Jack Kelly, a retired air force officer, and his superior, Conrad Lautenbacher, the head of NOAA, was a retired vice-admiral, gives the organization a military feel, a bureaucratic advantage in a moment of conservative ascendancy. The weather service is returning to its roots in the Army Signal Service.

This change rubs some the wrong way, a culture shock for an agency that had previously been managed by scientists. One forecaster persisted in referring to General Kelly as "General Halftrack," after the ineffectual leader in *Beetle Bailey*. Not all forecasters satirize General Kelly's leadership, however. The organization with its quasi-academic tone was a troubled agency prior to his selection. The NWS was not trusted by the

White House and Congress. One forecaster explained about a former director, "He was a scientist. We were using pencils down to the nub. We need an advocate to go to Congress. The organization has always been run by scientists, and I think that has hurt us" (Field notes). Over time the agency was being starved. General Kelly with a background in air force meteorology was asked to write a report evaluating the agency, which was facing deep budget cuts, layoffs, and office closures. The report, calling for additional funding, was so favorably received by policymakers that Kelly was given two years to turn the agency around. His success, at least from a bureaucratic perspective, has been notable, and the agency is frequently cited by the Bush administration as a success, a case of effective and measurable ("objective") accountability. A change in organizational tone is recognized:

> Certainly now [there is a military tone]. Not before now. It certainly was the opposite before, when Joe Friday was in charge. Now we've got the general in charge. . . . I think there's more of a depersonalized approach. It's more of orders, it's more top-down type of thing, and that's not entirely bad. I mean you get things done that way. You can make decisions much quicker. They may be wrong decisions, but you take that risk because you want to move ahead. That's the way the military is. Make a decision, do not be indecisive, do it now, make it and if it's wrong, it's wrong. . . . Under the old regime, you studied and studied and maybe you studied it to death. You just took way too long. . . . The negative to this [military system] is that when you come down with this kind of an approach is that a lot of people shut up. . . . I don't like that, but that's the way it is. (Interview)

: : :

Jerry, an administrator at the National Centers for Environmental Prediction, tells me that he is frustrated with the change in the organization. He sees the group as being downgraded to a level equivalent with a forecast office, even though he sees development as being more significant than operations, but in the military there is no separation between development and operations. He suggests that his group is more like a university science department than a military unit. In his group, "people don't like to be told what to do. It's like herding cats." (Field notes)

With General Kelly as director the ongoing tension between "base and gown" was salient. The agency has traditionally hired both former military forecasters and budding academics. The former are comfortable with a general in command, the latter would prefer a dean. In addition to personnel choices, the rotating shift system is tied to the demands of military life, building a band of brothers; the research centers with their academic hours are modeled on the university. Finally, the largely male workstaff (in part an outcome of rotating shifts) gives local offices a masculinist, if not military, feel; research centers are more diverse, both in gender and national origin.[11] While gender has not been a central theme of my organizational analysis, the NWS is structured so that employment is easier for men than women, given the structure of family life. Men are seen as fitting more easily into the 24/7 demands and are more closely tied to the agency's self-image; it is said that women are more likely to resign or to take an administrative position rather than continue to work shifts.

The debate over whether the NWS—and government science in general—is academy or military is played out against the recognition that it is bureaucracy. This view suggests that neither operations nor knowledge is the goal, but institutional survival is primary. While the strong form of this belief is not much heard within the agency, occasionally its echoes are evident, and certainly workers are sensitive to this concern. On one occasion I was told that a meeting that had been scheduled for Lake Tahoe was moved to Portland, Oregon. Lake Tahoe wouldn't look right. Some employees explained that they considered themselves overpaid, taking their political beliefs to heart in a way rarely found in the academy.[12]

The greatest complaint about headquarters, one that echoes the rhetoric of President Bush, is the absence of accountability. Who controls them, and to what end? This is not the same as arguing for the accountability implicit in verification statistics but for a sense that someone should be responsible for organizational policies and practices, avoiding what one frustrated forecaster described as "this multiple tier of incompetence" (Field notes). At local offices this often means that superiors are not held accountable:

> Ralph informs me that the NWS cannot determine which offices are successful and how to improve those that are not or which individuals require additional training. He explains that local problems could be used for education, as opposed to evaluation.

> He concludes that "federal people are not big on accountability. . . . It was all chaos. There isn't the discipline. . . . There are a lot of people who do not want to [change] because they would be accountable." (Field notes)
>
> : : :
>
> Stan explains how the IFPS system has been misused, "In Washington generally, they say we don't want to get involved with it. There's no one supporting it. It's amazing the amount of creative weaseling that people use to evade the rules. . . . Our regional director says if we don't do this, there won't be a weather service, but if this is all there is, what are we doing? If this was commercial software, you couldn't sell it, but damn the torpedoes, we're going to do it. They make a commitment to Congress, but they can't do it." (Field notes)

The issue is not whether these critics of accountability are accurate, but rather this rhetoric uses a third model of the organization—not a university department, not a military unit, but a bureaucratic agency with everyone attempting to avoid responsibility and to protect their turf.

Centralization or Local Autonomy. A core concern within this organization, and indeed all organizations, is how much centralized control is desirable. To what extent should the organization be disciplined by a central authority.[13] Consistency and order have benefits, particularly in an agency that bills itself as the "no surprise weather service." Yet, such order may depress the creativity, the autonomy, and the awareness of local needs that are necessary for effective response to unpredictability.

Oscillation in attitudes toward centralization occurred throughout the history of the agency. What is the proper size and scope of responsibility for a local office? With the creation of a set of 122 local offices (replacing the larger state-based forecast offices and the 230 smaller offices with a single meteorologist on duty without forecast responsibilities) the movement is toward forecast decentralization.[14] This major change in the structure of the agency occurred during 1993–99 and had been fully operational for two to three years at the time of my research. The new offices with many younger forecasters were still gaining collective skill in forecasting and in learning about their new communities.[15] They lacked the confident authority found at older offices.

In contrast to the decentralized model that developed, at one time some believed that the National Weather Service should maintain a single forecast center in Kansas City. The building would have four areas for the regions of the country, and presumably this spatial contiguity would allow the forecasts to be coordinated nationally. As one said, "Forecasts could fit together. Someone would put their stamp on it"—something that is not true today with local forecasts. This system, appealing as it might be, would eliminate local knowledge and community service. Rumblings are now heard of the same desire with the installation of the IFPS gridded database. This system encourages coordination among offices, and so a regional center might mitigate the problem of fitting forecasts together. In addition, as forecast models become increasingly sophisticated, and if public forecasters are unable to improve on these forecasts through their local knowledge, the idea may gain traction, even if outposts remain for severe weather warnings. As bureaucratic fads run in cycles, the future may bring centralization. These are ultimately philosophical choices emphasizing some aspects of the agency's responsibility. The weather service is likely to continue to oscillate between centralization with its appeal in control and consistency and decentralization with its appeal in autonomy and service. That the National Weather Service now requires a single template for local webpages, creating a distinctive "corporate image" to compete with the Weather Channel or AccuWeather, suggests that centralization has not lost its appeal.

As is often the case, the solution may be rhetorical. One administrator, after a talk by General Kelly, reported to his colleagues, "We're not standardizing things, we are making things more consistent. When Kelly is around, we do not standardize" (Field notes). The dilemma of centralization, thus, becomes handled by a linguistic device. Consistency, assuming interorganization coordination, fits the desire for local autonomy in a way that standardization, implying inflexible centralization, undercuts. Perhaps the outcome is similar, but rhetoric permits particular values to be set as the criteria for future decisions.

Regional Politics. One of the aspects that I found most startling about the NWS was that it was organized by region. I couldn't imagine the U.S. Postal Service being so structured. Perhaps I shouldn't have been surprised because the U.S. Customs Service is organized by region and until 1998 the Internal Revenue Service was as well.

Recently there has been a push to nationalize the NWS (and other agencies) to help with coordination. I was told that in the past "regions ran the show." A forecaster asked sarcastically, "Why don't they call it the Regional Weather Service" (Field notes). Regions still remain powerful. Under this system, the offices (in the continental United States) are divided into four regions, each with considerable decision-making authority: eastern, southern, central, and western (there is also an Alaska region and a Pacific region, but these are smaller and were not discussed by informants). Many meetings of administrative staff are organized by region, and awards and commendations to the staff are signed by the regional director, not by an administrator in Washington. To some extent this structure results from different weather problems: winter storms in the East, hurricanes in the South, tornadoes in the central region, and fires in the West (because of drought and winds, fire is treated as a weather problem, rather than a problem of land management). A heavy snow in the southern region would be 4 inches or more, and in the central region it would require 6 inches to be so labeled. Yet, the regional structure placed Oklahoma City in the same region as Miami, and Cheyenne in the same region as Louisville. Further, each region not only had its own weather challenges but its own organizational style. Regions were perceived as having distinctive cultures and were organizational fiefdoms:

> Regions have significantly different customer bases and significantly different attitudes toward what they require out of their forecast offices, and, as a result, the programs for different regions are much different. The eastern region tends to be very much of a population center. There's a lot more public outreach; there's a lot more contact with the people that they have there. The central region tends to be very focused on severe weather. . . . The western region tends to be focused much more on differences due to the elevation changes there. . . . And southern region tends to be more hierarchical in its structure, so it has more focus on the process of getting things done than on any particular segment of their weather. A lot of it comes down to the personalities of the management. (Interview)

His explanation mixes the meteorological and the social. Regions become typified, whether justly or not. One forecaster, speaking of the western region, commented, "That's a whole different weather service out there," referring to their tradition of bureaucratic independence. A second commented that, because of the population centers and media outlets there, the Eastern region was given special perks. A third commented that the southern region was particularly supportive of its field staff, resisting the push to cut personnel, and the central region had the reputation for being more financially cautious than the rest. The differences are such that the regions develop their own computer programs, not shared with other regions. The forecaster quoted above explained that this organizational diversity even affects the wording of forecasts:

> People in the East are much more direct in their forecasts. There's a lot of shading that goes on in the Midwest forecasts, especially if you read synopses or products associated with warnings. There is a lot more "if everything works out, this is what will happen." There's a lot more couching that goes on. In the eastern region it's more much direct. "This is what's going to happen and this is why" and I think that's a representation of the different personality traits between East Coast general public and Midwest public. (Interview)

Forecasters, even though they move from office to office, typically transfer within the same region. Forecasters who switch regions comment on the difficulty in adjusting to the new culture. In some offices hiring within a region is preferred in that, given the different regional cultures, within-region hiring means less culture shock for the new forecaster and less disruption for the office.

The NWS operates as a segmented bureaucracy. That the authority of the region is becoming eroded indicates that the debate is being played out between centralization directed from Washington versus local control, with the mediating effects of regional headquarters less important. The hybrid form of organization that once characterized the service is losing ground in the battle between uniformity and autonomy.

A Bureaucratic World. Whatever the claims of science, the National Weather Service is a bureaucracy. It operates under the umbrella of the Department of Commerce and the United States government. President Bush has the authority to mandate policy for the agency, to set its tone, and to select its staff. When Washington says jump, local offices at least hop. Individual meteorologists may chafe under the decisions of distant administrators.[16] The goal is to create a "national weather"[17]—not a "Chicago weather" or "Flowerland Weather"—but in so doing issues of authority and budgetary resources rise to the fore.[18]

Rules made from headquarters in a bureaucratic regime may constrain or channel meteorological work. During my observation, headquarters pushed for offices to increase coordination. As a result, forecasters were told that they had to document the calls they made to other offices. After a month when there had been only five calls (five *documented* calls), forecasters were instructed to call more frequently. This meant that if the person one was trying to reach was not available but called back, it had to be specially noted as a call by the original caller and not by the one who returned the call. To be called receives no bureaucratic credit. Because of this organizational demand, more coordination calls were made, although there is no evidence this produced better forecasts.

On another occasion a staff member needed to visit weather recording sites ten minutes from his home (an hour from the office). The problem was that the van that held the equipment could not remain at a private home overnight, and so the next day, this technician had to drive to work to retrieve the van, spending an extra two hours. Even though the rule preventing using government vehicles for private business is surely reasonable, the lack of discretion is problematic in practice. One administrator told me that he was required to spend 1.5 percent of his time dealing with issues of diversity. He asked me what I felt as a sociologist he might do to promote diversity and then counted our talk toward his administrative requirement.

A potent bureaucratic threat is an evaluation by the Inspector General's office, a full-fledged, critical assessment of office operations. The staff of the Chicago office, knowing how far they bent the rules, was fearful that they might be a target of one of these visits and were greatly relieved that they were not selected:

> HARRY: The IG is not coming here. They are going to Minneapolis.
> RANDY: Maybe we can do something to get them here?
> HARRY: Don't joke about that.
> GEORGE: Thank goodness. The office is never the same after they're done. They set people against each other.
> VIC: These guys make the Gestapo look like Brownie Troop 27. (Field notes)

That review found violations of security procedures, lack of adequate control of VISA cards,[19] unsecured sensitive items, and other complaints about the administration. Staff at the two offices at which I heard these discussions felt that they had the same violations, even though they did not see these violations as troubling. But they did feel threatened; one of the administrative assistants joked, "I don't want people coming back to me and saying, 'Bye-bye. It's Leavenworth for you.' I'm not going down by myself." While she joked about doing time, such evaluations could shape careers and affect office reputations.

Central administration engages in an ongoing struggle to formalize and standardize local practices. Field offices have practical problems that are not always easily amenable to rigid policies. Local cultures must battle against central authority to permit flexible responses to local challenges.

Oversight and Undercutting. Weather services are organized by the nation state. There is a Canadian service, an Australian service, a Japanese service, and so forth. This provides the state an entrée into the control of weather reporting. Although federal agencies have some measure of internal autonomy, they are overseen by the executive and legislative branches with their own interests. Meteorologists collaborate with those who control institutional resources to mobilize support for their activities, creating a weather lobby.[20] When confronting an agency with a national scope, with offices in nearly a third of the congressional districts in virtually every state, politicians are tempted to claim a piece of the action. Even if this stops short of scandal, it shapes agency decisions. The map, meteorological and political, becomes a game board on which decisions about the future of the weather service are played.[21]

I was informed that for some time there has been a push to split the southern region between the eastern and central regions, creating a three-region solution, saving expense. This plan has been blocked by southern politicians, particularly those in Texas, the headquarters of the southern region. The decision to move the Storm Prediction Center from Kansas City to Norman, Oklahoma, was described as a battle of "Oklahoma versus Bob Dole." The Chicago office, located in then–House Speaker Dennis Hastert's home district, received several new radio transmitters through his intervention. Alaska Senator Ted Stevens managed to get several new wind observation stations for his state, rather than the Midwest, because NWS needed his support for their budget. Some offices ("political offices") such as those in Key West and in Huntsville exist because of congressional pressure. One Wisconsin congressman felt that the warning area of the La Crosse office was too small, and so a county was moved from the Milwaukee office to satisfy him. I was told that even the criterion for hail (3/4-inch hail) as severe weather is political, allowing farmers to receive insurance benefits for crop loss, even though little damage is done by hail this size to persons or buildings.

The agency knows how to play this political game. Sometimes damage surveys after severe weather become political in that these bureaucratic decisions justify local disaster funding. Projects to improve technology are also constructed politically. The most salient example was the pressure on Congress to provide Doppler WSR-88D radars to each local office. While the radar has improved severe weather forecasting, I was told privately that it was "oversold": "they sold Congress a bill of goods," claiming more power to see tornadoes than the radar could provide (Field notes). The presence of this instrument increased the pressure to disseminate warnings, because "we need to justify our funding" with "underlying political reasons that drive this" (Field notes). Despite the statistics (more warnings and more confirmed severe weather), there is no meteorological reason that the United States should be facing a tornadic epidemic. There is a political reason. The weather service must demonstrate that radar funding was well spent and that the agency is accountable. This is a direct response to the hot air and turbulence that emanates from Washington.

Legitimating the Future

Science does not only speak to itself, it has audiences that struggle to hear. These are the consumers of its claims of the future. Along with data, theory, and experience, for a prediction to be believed, the occupation and organization must create credibility in light of the needs of its audiences. Futurework depends upon the standing of the claimsmaker, legitimating the prediction. The future does not stand on its own but requires credible social support. In this sense, forecasting is political, striving for allies that justify its claim to expertise and preventing predictions from being questioned. If scientists are experts, accepting their predictions becomes the default, unless there are strong and compelling reasons for doubting specific claims.

To understand the interests of audiences I focus on three overlapping audiences of operational meteorology: the general public, the private sector, and the media, each with links to the others. Each audience makes demands of the NWS for information that it can use and by which it can profit. This agency that describes itself as "customer-focused" aspires to present their work to meet these demands.[22] In this, the agency desires to be seen as *trustworthy*.

The presence of multiple audiences with separate interests in using meteorological information emphasizes that *weather*—and certainly the forecasts of weather—constitutes a *boundary object*, used by communities in different ways. Meteorological claims have varied meanings and implications in different social worlds. As Geoffrey Bowker and Susan Leigh Star emphasize, "Boundary objects are those objects that both inhabit several communities of practice and satisfy the informational requirements of each of them."[23] The thunderstorm constitutes a different kind of symbolic object for the media broadcaster, the farmer, the truck driver, the toddler, the operational forecaster, and the academic meteorologist. Their responses to thunderstorms vary widely.

In the Public Eye

The public is often profoundly ambivalent about the competence and efficiency of weather forecasters. On the one hand, meteorologists seem an esoteric priesthood explaining the mysteries of the universe, an image enhanced by jargon and sophisticated technical images. By speaking in arcane language in the cloak of science, predictive workers gain stature. On the other hand, forecasters are a source of amusement, displaying endearing incompetence. Inaccurate predictions are recalled in humorous accounts. The public does not choose between these two images but draws on both as appropriate. Combining the two images puts weather increasingly in the realm of entertainment and, as Marita Sturken points out, when well presented, can even contribute to our regional identity.[24] The skies determine who we are.

Despite the jokes and the awe, we follow weather reports closely and treat them seriously. In general, we trust weather reports, in part, because we have little choice. If we wish to plan our day, these forecasts are all that is available. In most cases our trust is rewarded. Science replaces the folk beliefs that once permitted meteorological confidence.[25] The NWS and their media translators emphasize the predictive value of these claims, a stance that serves both institutions well, as it also does for medical, nutritional, and financial news.

Media outlets have input into the forecast that they provide. With the exception of those who rely upon the NWS webpages or weather radio, for most of the public the weather information that they receive is mediated—shaped, selected, massaged, and visualized by media sources. While broadcasters select what they report, they lack direct access to the information of government forecasters. In many cases, what is available through the media is a translation, with some tweaking, of what government meteorologists have disseminated in their forecasts and forecast discussions. As Bruno Latour suggests, however, official weather forecasts are often swamped by a tide of voices, public and private.[26] Latour argues that the voices of operational forecasters are powerful at the end, determining what weather *was,* not what it *will be*,[27] but he doesn't emphasize that their claims and data begin the conversation from which others can speculate and provide the warrant for that discussion. Government forecasters, while sometimes hidden, provide *talking points* for local knowledge.

When the public demands information, forecasters may receive excess honor. Their audience may inflate the hopeful claim, proffered by their employer, that these workers have techniques to reveal the future. One forecaster noted, "The general public gives us a lot more credit for science than is really going on. . . . Some people call up for a specific forecast two or three months down the road" (Field notes). I was told that:

> They think we have more ability than we do. I'm sure you've heard the stories about people calling today asking, "My daughter's going to have an outdoor wedding at 3:00 p.m. on June 21. What's the weather going to be?" Sometimes it's easier just to tell them, low to mid-70s, 30 percent chance of thunderstorms, just because that's what the climatology says, instead of arguing with them for two hours trying to explain to them why you really can't tell them what it's going to be at 3:00 p.m., June 21, 2002 in their backyard six months away. (Interview)

A third forecaster relates, "Most people would be pretty surprised by how many mysteries there are in meteorology. People assume that we got it figured out." He adds that the public feels that "If we missed it, it's because we're lazy. We have our feet up" (Field notes). They are tripped by the confidence in the futurework that their agency sponsors. The presence of *numbers* in the forecast is deeply comforting, because their inclusion proclaims that science is at work. These numbers are not only included but are emphasized to suggest that the NWS is more scientific than media and private companies.[28] They are staking a claim to knowing; the precision justifies their authority to forecast.

This assumption of expertise cuts two ways. If science can tell all, the blame for errors must reside with the human forecaster.[29] Faith looking forward becomes discontent in retrospect. The public assumes that these government workers, and their media counterparts, must just be going through the motions. The belief in the certainty of science suggests that warnings should be perfect. The public must be trained in the inevitability of uncertainty,[30] but this is not something that forecasters wish to emphasize until they must account for their errors, which happens all too frequently. The criticisms and sarcasm aimed at meteorologists are legion, widely felt by forecasters:

> I was standing at the check-out line at K-mart. I was on a midnight shift, and my forecast was that there was a decent snow that morning, snow showers. I think that it accumulated an inch or two, and the forecast was for it to be sunny in the afternoon. I was standing in line, and there were a couple of women ahead of me, and they were mentioning how bad the weather was outside, and they said [sarcastically], "Yeah, the forecast was for it to become sunny and it's not going to happen. It's just so nasty out there," and I felt like saying, "I wrote that forecast." . . . You feel like saying "I'm they." (Interview)
>
> : : :
>
> We're still the butt of a lot of jokes. There's no question about that, and I think that's something that you get used to after a while. . . . Even my dad used to tease me about this. That "it's the only profession where you get paid for being wrong." (Interview)

One forecaster indicated that a member of his family commented in a similar vein, "I wish I would have a job where I could be wrong 90 percent of the time" (Field notes), and, as I was completing this manuscript, an op-ed in the pages of the *Chicago Tribune* by humorist Barry Gottlieb complained that: "weather forecasters are exempt from the normal guidelines of job performance . . . it's not right that my horoscope is more accurate than the three-day forecast."[31] The meteorologist must cope with a public that doesn't understand the problems of forecasting. In this, they are like so many workers who feel underappreciated and misunderstood.

A Helping Profession. Meteorologists, at least when on their best behavior, see their task as helping their fellow citizens. They provide a "public service." Knowing the future can help people adjust their routines and improve their quality of life. Successfully predicting severe weather can prevent injuries if the worst happens. One forecaster explained that he and his colleagues prefer writing the public forecast, because it is used by more people (Field notes). Another noted as a storm approached, "We're just here to help people out" (Field notes). And, as was suggested to me on several occasions, "we go home and we have to shovel [snow] too, so we see it from both sides" (Field notes). As consumers of weather information themselves, they care about blown forecasts.

Some forecasters complain that an emphasis on scientific precision distracts from their mission as public servants, diverting their attention from their audiences:

> I think you need to take all this stuff that should tell you whether you need to wear a coat, and whether to take an umbrella. . . . I think too often we forget what we are doing this for. Who's using it. We get too fascinated by the meteorology. . . . We don't think about who we are talking to and what the people want, and we never bother to ask. (Field notes)

This led to the forecaster keeping counties in the metropolitan Chicago area in a single zone, figuring that commuters move from county to county and wish to rely upon a single forecast. His concern derives from a belief in how people utilize his forecasts, even if smaller zones might be more meteorologically precise. Likewise, even if drizzle is meteorologically mundane, it is central to the quality of life. A forecaster who didn't announce that cloudiness could become short periods of light rain would find an irritated public. By emphasizing their public service, as opposed to their science, they respond to the challenge of private companies, such as the Weather Channel, that claim service as their mandate. When weather service personnel emphasize public safety, they embrace their similarity to the police, emergency personnel, and other essential governmental services.

The weather and the forecast are boundary objects, understood differently by different audiences. What meteorologists communicate may be taken in unintended ways. The continuing goal is to choose words that speak to a public and bolster professional competence. Forecasters as popularizers are the bulwark of the scientific state.

A Public Nuisance. The public is not neutral about the weather. Nothing can make people feel more *alive* and *embodied* than a beautiful spring day after a cold, chill, gray winter. Schoolchildren *pray* for snow. Others relocate from areas in which floods or hurricanes are a perpetual source of anxiety. Guides organize tours to permit tourists to experience the thrill of chasing tornadoes. Or as John Seabrook reflects about blizzards, "Unlike hurricanes and tornadoes, which provoke as much dread as wonder, a blizzard is anticipated mostly with pleasure, at least on the East Coast."[32] Many people wish to experience something memorable, an event that becomes a personal narration. As one forecaster recognized, "They want to experience something greater than average. If they have a straight line wind event [*only* a thunderstorm], people sometimes get upset because it was not called a tornado" (Field notes). People have intuitive and emotional connections with those atmospheric conditions that swaddle them, and see these conditions as the stuff of talk and of collective memory.

As a result, people are personally invested in weather. While government meteorologists lack groupies, there are people ("weather nuts" or "the weather service's little fan club") who avidly follow the weather, keep personal records, and call the forecast office if they believe that a forecast is in error. Like medical patients and savvy investors, the weather-invested public feels that it has the *right* to information.[33] By mid-December the Chicago office begins to field calls asking if there will be a white Christmas; during the summer a man called asking if they have a "barbeque alert" (Field notes).

The potential—if not quite the reality—of public harassment was felt to be such a problem that some forecasters use code numbers to identify themselves on their forecasts, not signing their names. The idea of a meteorological stalker might seem implausible, but from the hot seat it feels all too real.[34] Yet, this "public harassment" is believed, and one recounted a rumor that an irate citizen tracked down a forecaster at home to complain about a blown forecast. As one told me, referring to working in a large city, "there are a lot of lunatics out there" (Field notes).

Some people are inordinately frightened of the weather—storm phobics[35] and a few call their local forecast office whenever severe weather threatens, confident in the expertise of their forecasters. The Chicago office had three regular callers, two were said to have cognitive impairent. One apparently believed that the government controlled the weather (a worker joked, "He's got to have an FBI profile"). The staff was fond of these callers, asked each other if they had called when severe weather had been forecast, and did their best to accommodate them. The third caller was a storm phobic, a young professional they hoped to persuade to move to San Diego where the threat of severe weather was negligible or, if not that, to remain in his basement. As long as these callers limited their inquiries, they were a source of amused and sympathetic accounts. These calls and others like them are, however, sometimes called "kook calls" and, when intrusive, given other tasks, are quickly dispatched. Still, these three citizens may have been the most intimate, recurring contacts that the forecasters had with their public. In knowing the public through extreme and cranky instances, forecasters have challenges similar to those of the police, emergency room doctors, and customer service specialists.

Informing the public about the future is part of the forecaster's public service responsibility. Yet, this can be taken to excess. Meteorologists, like doctors, may be approached at parties and asked to provide a prognosis. Perhaps it is a compliment that the public believes that meteorological knowledge, like medical knowledge, is not left in the laboratory but is carried in the mind.

The Preferred Partner

There has been a push to privatize the functions of the National Weather Service, part of the same neoliberal agenda that has encouraged the devolution of government. But because of the investment needed for gathering data and the liability for severe weather, the agency remains. Nevertheless, the threat has left its mark. The weather service now accepts that it must justify its expertise through collaboration with private firms. A tension exists between the private and the public: how much scientific knowledge belongs to the public.[36] Big science can be held either publicly or privately.[37] Still, the linkages between the two are strong as public science depends on private industry for building equipment and private industry depends on the government for information from that infrastructure and for legal protections.[38]

Over a hundred private firms disseminate meteorological information.[39] Their relationship with the National Weather Service can occasionally be strained. In some domains, they compete with the NWS, although the weather service has privatized the responsibility for agricultural forecasts and does not provide personal forecasts for particular businesses.

Given that the likelihood of full privatization seems remote, I didn't hear much overt hostility toward private industry in local offices. Most forecasters see private firms as offering services that they do not provide, creating employment for their meteorological brethren. There is occasional resentment about salaries and reports of salaries,[40] which at the very highest levels are better than for government service, but at the lower levels are not as generous. Whereas the starting salary at the National Weather Service in 2002 was in the mid $30,000s, rising to $64,000, the comparable starting salary at a private firm I visited is about $20,000 rising to $40,000, significantly less than government work, although the variation among firms is wide. Some government forecasters referred to the salary at the private firms as "peanuts," the staff as "workhorses," and the attitude toward work as a "sweatshop mentality."

The resentment that exists derives from the belief that the Internet could permit the NWS to communicate directly with its public, competing with private industry, and yet the agency chooses to be "the preferred partner" for private firms. The development of a detailed forecast in the form of a database serves the interest of private firms, who can take this government database to provide spot forecasts to contractors, utilities, ski lodges, suburbs, snow-plow services, commercial growers, sports teams, or candy companies.[41] One critic reflected about the transition of the weather service from public service to private support:

> First, [the private firms] don't want to be involved in the dirty part of the work. They don't want to be involved in warnings. They don't want to get the data. Take the cream, they leave us with the dirty work. We should be providing to the public. . . . I can't find the satisfaction in having the end product be someone else's product. . . . These businesses wouldn't survive without the government providing the information for free. (Field notes)

Critics argue that tax dollars permit private firms to survive; providing a database makes private firms more profitable. One former employee at AccuWeather, considered by some to be the weather service's archnemesis, explained that by providing the database to private companies, "It's like a steel company wanting you to give them the raw materials, and then they make the steel and get all the profits." With weather service data available on the Internet, a level playing field is created. In fact, some businesses might now access the NWS data directly instead of contracting through private weather companies, but unless they have confidence in their ability to read these grids, private companies provide added value. The unanticipated result now, some years after my observations, is that the public (and companies who might have paid for this information) may be sufficiently savvy that they can access these forecasts themselves. Databases can be read by anyone with the motivation to translate.[42] The irony that private firms demanded the very database that could in the future put these firms out of business is not lost on forecasters. For their part, private firms are pushing this data to be removed from the Internet, only to be available by subscription.

A public/private partnership is often endemic in science, despite the tensions. Yet, because operational meteorology is so public, the connection seems more troubling. Taxpayers provide data so that private companies can profit, which they could not do without the public weal. As a strategic matter in the context of American politics in the early twenty-first century, the agency's strategy is understandable, although the ultimate effects may not be what anyone expects, as Everyman can become his own meteorologist.

Mediated Weather

Societies receive the weather news they demand. As Jack Fishman and Robert Kalish have described, the image of the media forecaster has shifted since the beginning of television.[43] We have watched clowns, sweet young things, silly old men, and handsome scientists. With increasing interest in weather and sophistication in meteorological forecasts, weather girls have been nudged offstage by broadcast meteorologists of both genders. New images—the weather map as a cultural text[44]—are presented, and viewers are socialized to the expertise that stands behind their presentation. Increasing numbers of television stations have their forecasters obtain "seals of approval" from the American Meteorological Society. From 1959 to 2002, over 1300 seals were awarded. These broadcasters are performers, but at the same time they are—or play—scientists, infusing a dose of faith in the future in a secular age. Their demeanor contributes to the trust that they are granted.[45] As New York artist Corin Hewitt commented, describing his statue of media forecaster Willard Scott, "they fulfill a very unusual role that transects both science and religious impulses."[46]

In many markets, one broadcast meteorologist gains special standing, and is revered for his or her expertise. In Chicago the "weather guru" is Tom Skilling, who forecasts on television and edits the weather page for the *Chicago Tribune*. The operational forecasters I interviewed were unanimous that he was the most widely respected forecaster in Chicago and his relationship with the agency's operational forecasters is strong. His link to the National Weather Service office is cemented by the fact that he hires retired forecasters. Even if stations have their weather information packaged by a private firm, such as WSI or AccuWeather, most broadcasters cooperate with the local NWS office, and on occasion dispatch a crew for a feature. Both parties find that connection desirable. The Chicago office sponsors an annual meeting to explain their procedures to the media. Surprisingly, government forecasters claim not to resent the high salaries of their media counterparts.

Many operational forecasters could not imagine themselves as media personalities and enjoy the security, stability, and anonymity of government work.

Currently the most important source of weather information for much of the public is the Weather Channel,[47] reaching over 95 percent of cable households in the United States. Perhaps the most significant blow to the relationship between operational meteorologists and their media colleagues involved TWC's change in presenting the local forecast. Forecasters are used to having their predictions altered by radio and television personalities, but the one place in which they—and their public—could actually see their words was on the local forecasts on the Weather Channel ("Local on the 8s"). Every ten minutes the Weather Channel would display the forecast for the next two days as written by local forecasters. But beginning April 30, 2002, the Weather Channel began creating their own texts—not a radical shift in content, or even in style, but no longer the same words. One irritated administrator told his staff: "It's discouraging. That has been the one place that our forecasters could look and see their forecast, and it's like someone hit you in the face" (Field notes). He had felt considerable satisfaction looking at the screen, thinking, "That's our product." It is a measure of the success of the Weather Channel that forecasters gain self-esteem from knowing that the public reads their forecast on this cable channel and that some felt an intense loss when those words were rewritten.[48] In the early years of the Weather Channel, the channel demanded changes in government forecasts,[49] but this resentment had been forgotten. This media outlet was their pipeline to the public.

Despite this blow to the forecaster's ego, erasing their words, most weather broadcasts stick closely to the forecast and data from the National Weather Service. David Laskin observed:

> The familiar tag line "Here's my exclusive Channel Q forecast" is a lie. "I hate this 'here's your forecast' business," says Eustis. "We all look at the same maps, the same numbers, the same progs. Yes, there can be different interpretations, but 90 percent of the time all the stations are so close that you wouldn't change your clothes or your activities no matter what stations you watch." As Ray Boylan, a weathercaster on WSOC-TV in Charlotte, North Carolina, puts it, "We are a delivery system for the NWS." All television weather is a set of variations on a theme. And that theme is written anew each day by the NWS.[50]

It is not that no differences exist between the forecasts on different channels, it is just that the variations depend on the forecasts that had been issued by the National Weather Service. As one meteorologist put it:

> You have people from TV and radio who take our products and value-add. The vast majority of creative forecasting is done by this office. Others reformat it. They rely heavily on what we put out. The weather service is the backbone for what they do. If you do away with the weather service, where do you think your models [would] come from? Where do you think your data comes from? How much does it cost to put up a satellite? This is the infrastructure that people use to make a living. That's all right as long as they recognize it. Don't slam the outfit, because we're the source of your profit, whether you know it or not. (Field notes)

Government forecasters salve their egos by distinguishing acting from science, relying on the image of entertainment: "they have some good people, but it's show business. As a rule, the real expertise in weather forecasting is at the weather service" (Field notes). Put another way, "The Weather Channel is show business. There are various levels of expertise on the Weather Channel. Those people have no responsibilities. They have no decisions. They don't have to make forecasts" (Field notes).

Operational meteorologists differentiate themselves from their higher-paid and more culturally visible media brethren, justifying their invisibility in the name of science. One's public self is not the basis of trust, but rather it is one's occupational competence and institutional legitimacy.

Weather as Drama. For media forecasting to succeed as cultural work, broadcasters must create dramatic narratives. This is not so different from the meteorological writing described in chapter 4. Media drama is constructed through a web of organizational and technological possibilities. Camera time on a half-hour news broadcast is a valued commodity. Weather forecasts are rarely given more than two or three minutes, unless the weather is the headline. Thus, broadcast meteorologists try to dramatize the weather, relying on narratives of threat, amusement, or aesthetics. One government forecaster suggested:

> They're concerned about the ratings. . . . People tend to remember the big events and forget the others so from their standpoint to stand out it can't be a good forecast to keep the job. You've

> got to hit the big ones or you've got to be ahead of the big ones, and that's news too. . . . We kind of get a kick out of team coverage of one-inch snow. They've got people scattered all over the city showing us wet interstates and they're trying to make a story. (Interview)

If wet pavement is what you have, dangerous highways become the news.

Often weather drama is centered on stories that evolve over days, notably hurricanes and snowstorms. These are events around which a television station can build an audience. Such weather events have a suitable narrative structure—a beginning, middle, and end, filling newscasts for up to a week. One forecaster told me about a media broadcaster who would go on the air on Monday and forecast a large storm for the weekend, "Someone asked him, 'Well, why do you go out on a limb, five days in advance on a big storm?' His response was, 'Well, if we say [it] on a Monday, people will tune in on Tuesday, on Wednesday, and Thursday'" (Interview). Not surprisingly media attention affects public consciousness.[51] Speaking of the hyped snowstorm of 2001, academic meteorologist Lee Grenci notes sardonically: "How in the world did forecasters in early March expect to pinpoint the area of heaviest snowfall two to three days in advance of the storm? Or perhaps more importantly, WHY did maps with predictions of specific snowfalls keep popping up on the screen like toll-free numbers to purchase miracle remedies for male-pattern baldness. Could the answer be television ratings?"[52]

My focus is not the media perspective per se, but how the linkage with the media affects government forecasting. As noted, the media take information from the NWS and transform it into drama, as was the case in the 1993 blizzard. As David Laskin describes:

> "This could be the biggest and meanest storm ever," Jacqueline Adams told the viewers of the CBS evening news. And there was Lou Uccellini [of the NWS], looking nervous, explaining that this storm was going to be a kind of winter hurricane, plunging barometric pressure to record lows along the East Coast. On the Weather Channel you'd think it was the eve of World War III. . . . "When those NWS guys start using terms like 'historic proportions,' 'record-breaking,' and 'extremely dangerous,' you know this one will come through," John Curley promised viewers on Washington's Channel 8. . . . And yet Uccellini admits that when people from the media swarmed into the NMC

> [National Meteorological Center] on Friday for a briefing, he was "sweating bullets" of anxiety that his office might have overhyped the storm. "CBS wanted me to call it the Storm of the Century. . . . But I refused to do it."[53]

The linkage between the NWS and the media is real; each depends on the other to justify their legitimacy.

Disagreements between the media and the weather service tend to occur on those days in which the weather service forecast lacks drama. These mundane forecasts fly in the face of the media's need to pitch a story. For instance, a local office predicted rain on a day on which all the television stations broadcast a forecast of snow. (It rained.) A forecaster told me, "We were the only holdouts. The TV likes to predict snow. If it were snow, it would be ten inches of snow" (Field notes). Forecasters joked about a broadcaster whose forecast for snow was "2 to 14 inches." Another said, "It's always better than what they [the media] predict." Stories are shared of TV forecasters who predict snow (wrongly, in these comic accounts) in the face of weather service denials, *guaranteeing* 6 to 10 inches (Field notes). When broadcasters are wrong, local offices must pick up the pieces, receiving irate calls. One forecast explained with some exasperation, "I can't help what you hear on the news. It's not going to rain or snow. It's going to be dry. . . . *We just tell it like it is*" (Field notes).

While relations are usually harmonious between the weather service staff and the media, resentments can occur. From the perspective of the government forecaster, trouble results when a media broadcaster competes against the NWS rather than against their media competition or creates a story by accusing the local office of some blunder. Needless to say, government forecasters are none too pleased with these strategies.

In the former, local stations sometimes proclaim their new equipment as being better or more sophisticated than that of the weather service. One station in Flowerland's coverage area was accused by forecasters of attacking "government radar" in favor of the station's "Triple Doppler Radar," touting its accuracy, not acknowledging the information received from the local weather service office. I was assured that the impressive term, "Triple Doppler Radar," was meteorologically meaningless: "It's an advertising tool. The capacity of that radar is severely limited because of the characteristics of the radar, but it's a selling point to get people to watch them" (Interview). This station is felt to be hostile to the weather service, wishing to take undeserved credit. Government forecasters remark whenever broadcasters at this station miss a forecast. A similar situation developed in another market, "He was not a fan of the weather service. He had his own radar, the TV station had some Doppler capability. So he would go out with Doppler alerts, and he would say, 'Well, the Weather Service hasn't done anything with this storm, but I'm going to issue a Doppler alert'" (Interview). Such strategies are not designed to win friends, only viewers. Gary England, a prominent broadcaster in Oklahoma City, created considerable animosity among staff at the weather service by broadcasting his own tornado warnings.[54] These broadcasters set themselves up as sources of expertise, despite the vast quantities of information gathered from the weather service. In this, they undercut the predictive authority and legitimacy of the weather service, a threat to government futurework.

Other forecasters are more directly confrontational, at least in the eyes of weather service employees, building reputations on institutional attacks. As one put it, "Some of them think they're like Mike Wallace of *60 Minutes* waiting for us to make a mistake so that they can cut us down" (Interview). The belief that the media only should fight each other, not government forecasters, was made clear by one forecaster who noted that, "It doesn't bother me when they poke at each other, but it bothers us when they drag us into it" (Field notes).

In Chicago, two television forecasters are criticized for gaining attention by attacking the weather service. I described how the mistaken snowfall amount, created by adding hourly snowfall amounts, produced a media onslaught with weather service incompetence becoming the story. Beleaguered forecasters claimed that the media was looking for trouble and that broadcasters were deliberately doubting their explanations in telephone calls described as "awfully ugly," "rude," and making "us seem that we didn't know what we were doing" (Field notes). A story about the weather became a story about government bungling.

Something similar occurred after another large snowstorm in which a broadcaster felt that the announced snowfall amounts were higher than they should have been, using a "political ruler" to help Chicago's Daley administration gain federal funds, hire new employees, and demonstrate competence at snow removal.

On a third occasion, a television station claimed that the office's heat advisory forecast was wrong, recalling the deadly heat wave of 1995. Such broadcasters were said to "like to stir things up," using the weather service as their target.[55] As a forecaster joked about one of these men, "He'd get better service from me if he didn't say who he was," and another said about one of his questions, "If he can't figure it out on his own, he shouldn't make the millions of dollars he does" (Field notes). I do not suggest that these broadcasters received poor service, but forecasters certainly didn't go the extra mile to be helpful.

In these instances, the occupational community of meteorology is divided by the location of their work. When media meteorologists begin to compare their expertise favorably to that of the weather service or to see government meteorologists as bureaucrats and not scientists, resentment results. The idea that these are individuals with different jobs but common interests is undercut by the fact that in significant ways they do not have common interests at least in the short run. Government meteorologists need broadcasters to justify their expertise, but at times the spotlight of a capricious media is harsh and unfair.

A Public Science

Work can be viewed from the inside in light of small-group dynamics and culture or from the outside in light of structure and occupational culture. An organization composed of local groups involves independent competing managerial experiments. Situated above this diversity of action, however, broader structures hold local groups in harness in the face of local cultures. The fact that the National Weather Service is a government agency, a bureaucratic organization, is crucial. Offices can only set their own policy to the extent that their overseers ignore or do not object to these local variations. The battle over centralization is a pivotal tension in this agency, as in all bureaucratic domains. Flexibility becomes how an office can justify itself in the face of a demand for consistency. The establishment of regional centers, spin-up offices, and formal oversight shapes the debate. Standing outside of the agency are other agencies within the executive and legislative branches of government that have their own pressures and demands. These power centers must be managed, as there is a trade-off between autonomy and resources.

Outside the organization stands a set of audiences: the public, private businesses, and the mass media. To have their predictions accepted, forecasters must be more than correct; they must convince others through their rhetorical appeals that their forecasts are to be taken seriously as the most plausible account of atmospheric developments. To this end the National Weather Service strives to reveal consistency, science, and a commitment to public welfare.

Forecasters emphasize that they are engaged in public service. The problem is that except for their webpage and weather radio, their claims are translated—mediated—by others, and these others profit from their labors. Relations among organizations shape public science.

Even if these others, private firms and media outlets, provide the backing that permits the weather service to survive as a government agency, they also threaten to replace government service eventually. Currently the cost of gathering information and the legal environment prevent this privatization, but potential rivals threaten.

Because of their close connection with the public, the media are both allies and competitors. Their different structures can cause friction, and the needs of the media affect the products of the weather service including the timing and the formatting of forecasts. Simultaneously the media are the means through which the NWS becomes known, shaping public awareness. Work is linked to institutional relations, altered through decisions made over time. Whether the weather service would prefer autonomy (and, of course, it would), at present that is not to be. Its role is to serve audiences, balancing their needs as professionals with the demands of these clients. These audiences transform trained meteorologists into civic actors and make their technical expertise a public science.

7 Weather Wise

We are swaddled in weather: sun and rain, wind and snow. Atmospheric events organize our lives, affect our economy, contribute to social problems, and determine our quality of life. Were we to have a single, consistent weather—hot, cold, wet, dry—we could arrange our social relations to take those conditions into account. We would not demand a craft that predicts what the weather *will be*. What the weather *is* would suffice. We produce the future because we have an audience that demands it. We create a public science to satisfy this demand for surety. In most venues, the most striking thing about weather is its sharp changes and startling variability. One finds—from locale to locale—the bon mot, "if you don't like the weather, wait ten minutes." We require weather forecasters to be our guides to this uncertain hereafter. In this sense the meteorologist and the public co-produce the weather: we demand, they provide. Not only do forecasters mediate between the present and future, but they mediate between us and the future.

I base this account on a detailed description of National Weather Service offices in order to permit readers interested in weather and in work to appreciate the demands that are placed on these public scientists and the skills that they require. I paint a picture of life in weather service offices. But my concern is not to assess accuracy or competence. The embarrassing or disconcerting episodes that I describe could be duplicated in all agencies and all occupations. The informal procedures by which work gets done belong to all forms of work.

My view is wider than the skies. I examine how public scientists strive for autonomy and hope to serve according to their beliefs of what this public needs, while operating under a set of constraints imposed upon them through an organizational hierarchy and by their demanding audiences. In this, they, like other workers assigned the task of prognosis, must envision and then colonize the future. To achieve this, workers and their machines produce a record of past and present that permits them to narrate the future plausibly.

That weather forecasters think of themselves as scientists (in some regards) does not change the reality that meteorology is similar to many work domains. Of course, these workers are not producing widgets; they produce knowledge. They create this information routinely and rely on sources outside their workspace. Operational meteorologists do not work on an assembly line, but some of the repetitive demands of factory work are evident in their shifts and in their products. Forecasters share the expertise of the research scientist, but what they know is constrained by the *guidance* provided to them. Their production is bureaucratized, and their shopfloor is a governmental office in which spatial privacy is surrendered to oversight by administrators.

Throughout my account of the work of operational meteorologists, several themes appear repeatedly. Specifically, I address the importance of *local culture* as a means of structuring work, the centrality of the *authority to know*, the *linkage of past and future*, the *right to communicate*, and the perception of the *autonomy of nature*.

What is particularly striking about the organization of the National Weather Service—although the agency is not unique in this—is the division of production into numerous parallel but linked units. The agency operates through a reticulated or networked structure. Even though the products might be seen to be reasonably similar, the conditions under which these products are generated are distinct, exemplifying both the power and the limits of group cultures and local organization. Communal norms encourage the development of a moral order on the shopfloor, establishing boundaries of autonomy and control.

Second, these are workers who are expected to have expertise. Their position involves more than technical skills; it includes knowledge that can be applied to solving problems. Their work location provides them with the authority to know, but they must demonstrate it by presenting an occupational identity. These are meteorologists, scientists, experts. They ask us to trust the information that they provide, permitting us to navigate our world successfully. This structures our action, preventing a fearsome or vexing uncertainty. At times this knowledge can be a matter of life and death, of health, and of the protection of property.

Data about atmospheric patterns depend upon gathering and assessing information. These workers have the authority to interpret these data, taking millions of bits of data and distilling them into a telegraphic assessment. Like other public scientists, they are knowledge funnels. We accept their selection because these workers have been credentialed: in this case by their B.S. degrees in meteorology and by the organizational training for which the weather service vouches. This is then enriched by the expertise that comes from the experience, history, and collective memory that they acquire in the course of their careers.

This knowledge funneling reminds us of how tightly coiled past, present, and future are. Workers have no more direct connections with the past than they have with the future. Their personal knowledge of the present is too personal and local to be of much use for generalization unless they can claim more systematic access to that present through a network of machines. Their first task is to select those aspects of the past and the present that they believe will help them uncover the future. Both the past and the future must be created: the past through machines and memory, and the future through the extrapolation of that past.

Once forecasts have been decided upon through a routine process, evident several times a day and at moments of physical threat, these decisions must be shared. Forecasters believe that their words matter, and in this they are not so different from all occupations that depend upon canons of communication (medical staff, lawyers, restaurant servers, or realtors). Workers take pride in the precision of their phrasing and believe that others interpret them with equal care. In meteorology the technology is changing with potentially dramatic consequences for the self-image of the profession; the tradition has been to assemble words, the prized product. The words are the plain air through which members of the public are taught. Despite the desire to use words to communicate with one's primary audiences, literary technologies are often mediated. Words are bundled out into the world as orphans, hoping to find a sympathetic home. Because of the role of the media, this translation may be more evident in the case of meteorologists than others who have direct contact with their public. The reality that others muffle and reshape these words disrupts the direct linkage between the operational meteorologist and the public.

Meteorologists are no different from many Americans in accepting a belief in the *autonomy of nature*. By this I suggest that the environment is treated as an authentic reality: a domain that can be described but not bridled. "Nature" is a space that humans inhabit but do not control. The claim that we read the skies suggests that the skies are to be understood in their own terms. As with so many workers who examine natural conditions, these forecasters bow to this essential reality, feeling that their responsibility is to present this other world respectfully to members of ours.

Through this analysis, I emphasized the central role of group culture, knowledge claims, linkages between past and present, communicative credibility, and the authenticity of nature. These themes apply to all domains that rely upon routine interaction among workers, the creation of spheres of knowledge, the requirement to communicate outside of one's group, and the mediation of the social and the natural. Similar issues affecting work process are to be found in high school English departments, research laboratories at pharmaceutical companies, and offices of the Environmental Protection Agency. Workers participate in a politics by which they can establish (or lose) cultural authority.

Beyond these themes I draw upon concepts that enrich out understanding of how meteorology operates as a knowledge discipline, fighting to gain authority. Several concepts help us see that meteorology is similar to other domains. I address science, group, autonomy, futurework, truth, and audience. Through these concepts I treat meteorology as science, as work, as culture, and as oracle.

Science

Each discipline that huddles under the umbrella of science has peculiarities, a function of its social and cultural capital, and its links to external sources of power and resources. Perhaps meteorology does not have the prestige, the organizational authority, the resources, or the elaborated theories of physics, but it is surely one of the physical sciences. Possibly its place as a public science works against it in the academic hierarchy. By speaking to a nonscientific public (in effect, dumbing down its language), meteorology becomes stigmatized, even while it hopes that some scientific cachet rubs off.

Not only is meteorology a public science, it is a *big science*. There is no place in meteorology for the isolated scientist. Forecasters are part of a network, both human and mechanical; the individual belongs to a team. The meteorologist, like the particle physicist, relies on external resources. Big sciences depend on state largess, and meteorologists reflect this linkage between political demands and disciplinary structure. The existence of national weather services involves more explicit control than is found in most other disciplines, but all collaborative sciences respond to the demands of political regimes.

In requiring an infrastructure and a technology meteorology is not so different from other sciences. As equipment evolves, what can be known is altered. While this new technology is claimed to serve us better, in some instances it extends uncertainty, increasing, for example, the margin of error in thermometers. These advances may not increase precision, but rather increase automation and the control of labor costs.

Technological advances encourage new organizational structures. The development of a new generation of radar is a case in point. This advanced equipment justified the decentralization of weather service offices. Offices were now sited according to the scope of the radar. The radar motivated the decision to create dozens of spin-up offices and led to hiring hundreds of newly minted meteorologists. The radar justified new offices. Once established these offices had to demonstrate their worth by issuing warnings, explaining the seeming increase in threatening weather as a result of institutional change.

Some argue that the establishment of meteorological databases may create demands for increased centralization, another structural choice. These databases require increased coordination as forecasts bleed into each other across the boundaries of office responsibility. Such a change could leave weather service offices with only their public service responsibilities, having excised locally created science.

Public scientists are agents of scientific authority, even as they also serve as agents of bureaucratic power. In the case of meteorology, severe weather may emphasize the science and routine forecasts the bureaucracy. These categories blur, however, as when during severe weather government agencies advise the public and state offices on their proper response, including activating emergency services, priming police, or demanding evacuation.

In observing restaurant cooks, I found that occupations do not easily fit categories of work. Cooks can be treated as artists, professionals, businessmen, or laborers,[1] depending on the task, the organization demands, and the audience. Operational meteorologists also have multiple identities; they belong to domains of science, administration, and public safety, selecting among images of academics, bureaucrats, and military officers.

Operational meteorology, like other forms of public science, is not strictly academic work, but at times staff borrow academic authority to justify their esoteric knowledge. At other moments, as processors of vital information, they may scorn the seemingly arcane research of professors. At what point do they fall within the domain of scientific research? When are they experts in public safety by virtue of the dangers they have experienced, and when are they merely reporters of technological readouts with little independent judgment?

The production of worded (and, now, graphic) forecasts reflects their scientific authority. Knowing the future invests forecasters with expertise. Severe weather is a critical moment at which the job alters its character, both for staff and for audiences. In this circumstance forecasters belong to an *activated organization*. They have the authority to direct citizens and state organizations, creating a chain of command. For this reason severe weather jolts staff from the mundane listlessness of announcing clear skies. In placing workers where the action is, emergencies emphasize their expertise, privileging their ability to preserve public safety. Through their pivotal position as guardians, they establish civic authority. While meteorologists have a unique set of occupational images, the reality that all occupations draw from a pool of images reminds us that work is a collection of disparate tasks.

The tension over where the occupation belongs in light of the standards of scientific production produces joking that plays off the uncertainty and ambiguity of their roles as scientists. When workers wish for, but also doubt, honorific authority, such humor is common. Put another way, boundary work appears in the clash between a hope for status and a fear that this status will not be granted. Every occupation faces threats to its autonomy and status; humor helps keep these threats at bay. A humorous remark allows threats to be defused through the claim of "only joking." Similar tension is defused by engineers, nurses, and laboratory technicians—those who are close enough to sniff the aroma of scientific authority but not quite close enough to taste it. But even groups that are treated as central to the mission of science use humor to defend themselves from their anxiety that others might question their position. Few domains of work fit an idealized model. Humor controls skepticism of one's work identity, but it also undercuts the status system and, perhaps, the honor of work.

Group

Work occurs within groups: a shared arena of known others, a local interaction order. It is in this social space that knowledge is formulated, evaluated, shaped, and shared. Knowledge originates locally. Sometimes it is diffused, becoming institutionalized, and sometimes it remains localized, tied to the group where it was produced.

All groups require structure, interaction, and culture. Each of these domains shapes the others. The Chicago office of the National Weather Service was a group with a lush and vibrant culture. It was a happy place for some and a difficult place for others. Yet, even though the doings of these workers were dramatic as cultural markers, there was nothing unique about the presence of cultural traditions, patterns of interaction, and local organization. For all groups, events need to be interpreted in light of local traditions. That the themes of the Chicago office have long been set emphasizes the power of a group culture. Even tragedy, personnel turnover, or technological change cannot erase long-standing traditions but can only shape them.

As people belong to several groups, they are influenced by their participation. We are each embedded in multiple social scenes. While we have some power to affect these scenes, the scenes determine how we identify ourselves and what behaviors and values we come to define as proper. Every group not only embraces tradition, but fights for authority through control of *local knowledge*. The group embraces a worldview, tied to its instrumental and expressive aims. Whether we examine families, clubs, teams, or workplaces, a group produces order and predictability by setting expectations for proper action.

Ultimately groups build on each other. Society is a network of groups. The individual offices of the National Weather Service legitimate the institutional power of the central administration, creating multiple variants of the organization, each sensitive to local conditions but dependent upon the authority of the whole. These groups may be linked in action or they may be models for each other (either positive models or as models from which a group differentiates itself). The connections among groups provides a basis for a structure that is built upon local interaction orders.[2]

Autonomy

The ability to set the terms of one's expertise is crucial to the realization of autonomy. Workers strive for the authority to know, demanding credibility. Yet, claiming personal or group expertise is often problematic within the context of a bureaucratic agency, particularly when the creation of information is outside the control of those who manipulate that data. When data are distributed in a hierarchy, recipients can produce forecasts only from the information that others have provided. That models and observations are funneled into a local forecast office means that the work products are heavily determined by external agents with their own agendas. Despite their claims of authority, these workers operate within institutional routines.

In addition to the hierarchical control of information is the problem of coordination, both spatial and temporal. With 122 offices, each adjoining perhaps five or six others, no forecaster is an island. Despite the boundaries of knowledge that forecasters may wish to draw, the need to create a single weather picture breeches these borders. Even if coordination is ignored, as it sometimes is, organizational rules limit the right of individuals to communicate as they please. Further, the forecast that a shift worker creates will be altered by his replacement. Any forecast will be changed as time, weather, and shifts pass. Perhaps this forecaster will return two shifts later, but the forecast may already have been altered in great or small ways. The forecast is fluid, creating a *composite autonomy*, rather than personal autonomy.

Through the demand for collaboration, knowledge production becomes collective, shifted from the control of any single knower. This is true for all organizations in which outposts must be coordinated to give consistent service. Few individuals have the right to make claims that others do not have authority to limit explicitly through direct forms of control or implicitly through gossip and private evaluations. Even professors, famously cloaked in robes of academic freedom, must confront administrators who can provide rewards and burdens, colleagues who may judge them harshly, and students who can vote with their feet.

The sociological problem is to draw boundaries of authority. These claims are always limited by a control structure that demands collective participation and accountability. Autonomy and control demand an equilibrium; the expansion of expertise is checked by those with interests in limiting those claims. A *politics of credibility*[3] affects our reliance on the claims of those who wish to persuade us. When actors establish boundaries in which their expertise goes unquestioned, their authority is enhanced.

Autonomy is a rhetorical claim as much as a behavioral reality. Workers have choices, but the acceptance of these choices is tied to the right to act that they have amassed. In organizations that depend on shared expertise, autonomy belongs to the group, more than to the individual. In the case of meteorology in which the obdurate reality of climate is close at hand and in which others are scheduled to take over the hot seat, the consequences of a strongly held belief in the face of contrary evidence of the skies can be severe.

Futurework

Some workers are given cultural and institutional responsibility to unmask the future. While many workers explain the past or help cope with the present, others must explain what is yet to come. While gazing back at the weather past is a matter of curiosity and explaining the current weather informs us what we could easily discover ourselves, the assignment of the meteorologist is what will be. Society depends on prognostication, allowing social systems to prepare for whatever shocks may transpire. Societies—perhaps the same society in different domains—use different styles of forecasting. We rely at times on art, religion, and science, even social science.

Prediction depends on data. Meteorologists, like physicians, petrologists, policymakers, pollsters, financial analysts, or even fortune-tellers, search out those data sources defined as legitimate and relevant. They choose numerous observation devices to gather data from which they believe extrapolations can usefully be built. An assumption that current conditions or recent trends will continue unchecked in a linear fashion is, however, insufficient. Practitioners need a theory that guides and justifies extrapolation. In the case of meteorology this involves fluid dynamics: the metaphor of the sky as an ocean of air is an image cribbed from physics. But whatever the discipline, some theoretical model to marshal and organize data is essential. Finally meteorologists require a history, their own and that of their colleagues. The future is historicized, explaining how the past has transpired, and what this reveals about current conditions. History is tied to the lived experiences of practitioners. Data, theory, and history permit the doing of futurework, transforming the present into a credible future. By proclaiming that they are examining the future, they rely upon their ability to construct and create both the past and the present. By deciding what to look for in the past and present, they build those times as well as creating the future. They create hindcasts as well as forecasts. Futurework is not only about the future, but about time that stretches forward and back.

Prediction demands more than the knowledge that is accessible to the forecaster. Central to the act of forecasting is belief, and belief depends on the ability of a social institution to make a case for its legitimacy, carving out a domain in which the rights to make claims should not be challenged. Science famously has gathered this authority, and organizations that can proclaim scientific credibility, such as the National Weather Service, have advantages. The backing and sponsorship of media institutions, creating forecasting credibility to gain audiences, help maintain this scientific authority.

Ultimately establishing the persuasive quality of the forecast is crucial. A forecast is a form of discourse that is backed by the authority of institutions. The trust that we place in physicians to diagnose and then to prepare us for those outcomes that they have predicted represents the strongest instance of how much we depend on forecasts to organize our lives and to put our bodily house in order. Policymakers—physicians of the public sphere—do for civil society what physicians do for bodies, although typically in policy domains many voices are heard, rather than the private, hushed conversations between doctor and patient. While few occupations are assigned the close predictions that burden meteorologists—the demand to know today about tomorrow—the desire for confidence in what will transpire is a common demand.

Truth

Truth is a social phenomenon. It has in the words of Steven Shapin a "social history."[4] Many workers are expected to be *right*, but what does this mean? Will we know it when we see it? The discussion of verification and other strategies for evaluating weather forecasts, watches, and warnings reminds us that the assertion of a truth claim is not the same as truth itself. And it is a challenge that is found throughout the work world.

Determining whether a prediction is correct depends on several factors. First, it rests on the desired level of precision. How right must it be? For instance, a forecaster can predict that the high temperature will be 53 degrees, but the thermometer may reveal a temperature of 53.1 degrees (if the markings permit such an exact reading). If we select degrees as our criteria (and we must choose Fahrenheit or Celsius, scales with different band widths), we round off to the closest number. We treat rounding error as *not error*. We also assume that the equipment represents how we standardize our evidence. It is this trust in equipment that assures us that we know what reality is, even while we know that a degree of error is built into the readings and cannot be excised. Accepting our embrace of a mechanically derived truth, we discuss whether a forecast that is off by one degree, two, three or more constitutes adequate precision. Only the most grumpy soul would complain about a prediction of 53 on a day in which the official reading was 54. As information gets translated, it sometimes becomes truncated. Forecasters typically provide the public with a range of temperatures and a likelihood of precipitation, but when the forecast is presented by the media often these ranges are compacted, making error more likely, although these media sources rarely trumpet their gaffes. Determining whether a prediction is right (and implicitly, how right) depends on institutional and collective choices.

These issues while applicable to meteorology pertain to other domains. How right must a medical prognosis be? Here physicians can sometimes avoid blame by ascribing the error in prognosis to the acts of the patient or the idiosyncratic responses of the patient's body rather than to an error of foresight. Medical care involves feedback loops in which the choices of the patient may mitigate or aggravate ailments. Something similar happens with the erroneous predictions of financial analysts who blame the psychology of the market for spoiling a perfect forecast.

Verification depends on whether an event is believed to have *actually* occurred. For instance, some tornadoes are ready to be photographed, but others are like falling trees in a forest. If not seen, they didn't happen. Verification statistics, managed by organizational needs and by chance observations, stand at some distance from ground truth, although verification may vary systematically over time or between offices or workers.

As an organizational matter, the establishment of truth is crucial for legitimacy. Even if we recognize those features that produce noise and error, some organizational standard is needed to suggest that the tasks at hand are being properly performed. This desire for confidence, found wherever accountability is demanded, makes the creation of truth a social process, but one that relies on the faith that we place on those—human or mechanical—who report the facts.

Audience

As students of science have emphasized, an object or act may be dramatically different depending on the standpoint from which it is viewed. These objects of multiple interpretation are labeled *boundary objects*. What we know depends upon the perspective from which we observe and the cultural filters used in our interpretations. Our demands as audience determine categories of meaning.

Weather has variable meanings for different people, groups and organizations. Operational meteorologists recognize these multiple audiences, shaping messages to provide those forms of knowledge that audiences desire. I watched forecasters cheering on tornadoes, testing their abilities at prediction, all the while their clients were deathly afraid that these divine curses would cause havoc. When they reported the threat they had to be aware of these fearful emotions. A tornado warning means that an emergency worker will have to work overtime, a child can stare at the skies with fascination, a rancher herds cattle, a mother is terrified for her family, a shop may close early, and a forecaster operates at a high pitch of excitement. On the basis of the philosophical tradition of pragmatism and the interactionist tradition of Anselm Strauss and Herbert Blumer, we recognize that an object gains meaning through the social responses of audiences.

It is not only a single object that carries meaning, however, but organizational relationships as well. The National Weather Service is a bureaucracy that is linked to political authority, public interest, private business, and media performance. Each audience has demands, even if they are not always able to state precisely and unambiguously what they are. Surely each audience wants accurate and timely information, but this has different meanings at different moments. A television station desires a long notice of blizzards and hurricanes to organize their programming to increase ratings. They need forecasts in the format that their performers can translate into language and images open to widespread understanding. Private meteorological businesses prefer raw data that they, and they alone, can mold. They want data that are not too accessible, for, if easily read, their value-added services would be diminished. The desire of private industry to have databases replace written forecasts serves this end. These firms become the translators of esoteric data.

The public for their part demands security, comfort, and certainty. The slogan of a "no-surprise weather service" markets to this desire. Students of risk suggest that the iconic goal of modernity is the creation of a *no-risk society*. The public may not be precisely certain from where the data derive, but the establishment of trust in weather claims and the faith that there is a government agency that watches over them provides confidence. This desire is transmitted to political authorities who strive to satisfy their public.

Audiences can limit the autonomy of occupations, even if they expand the occupational scope by raising the work's significance. While some occupations may be so esoteric that translators, such as science reporters or policy analysts, are necessary, few domains of work are fully cloaked from their public. Audiences are often present and vocal. In case of a public science these audiences constrain and channel the work, but they simultaneously demonstrate the importance of the work for the public sphere.

Authors of the Storm

Through the trust that we place in meteorologists, coupled with our skepticism, weather forecasting has become an essential feature of our modern no-risk society. By their predictions, these men and women permit us to order the complexities of our lives. Their claimed and actual abilities to read the future gives them influence and a place in our heart. For this reason the National Weather Service is one of the most used and most familiar of governmental agencies. Even those who are most skeptical of the power of the state are willing to accept the weather service as integral. Perhaps someday this agency, too, will be privatized, but until then the government must provide snug and secure information about the conditions of our lives.

Although it is not my goal to suggest detailed policy prescriptions, weather forecasting works best when staff recall that they are forecasting for people and not for land. Meteorologists should be aware of the social landscape of their area of responsibility. As forecasting is now done by drawing lines on a mapped grid, staff have become increasingly aware of the space for which they forecast. Yet, it is significant that these maps do not include population density or those institutions (hospitals, nuclear power plants, levee systems) that might generate special concerns.

Although the weather service forecasted the track of Hurricane Katrina accurately, the Federal Emergency Management Agency, unaware of the social context of New Orleans, was unable to respond to the needs of those citizens who were hit. One wonders whether the forecast for the hurricane would have been different had it been heading to open land. The hurricane might have been equally powerful as a meteorological event, but as a social matter it could not have been more different.

For meteorologists to be unaware of the census data of their area of responsibility, the infrastructure, or the human topography is dangerous. Surely one might reasonably desire a different warning in its urgency or timing for a major tornado that cuts through farms from one that slices through suburbs. My data suggest that forecasters know this intuitively, but they have not been trained in this form of decision-making. These issues seem tangential to the concerns of headquarters. To suggest that meteorologists recognize that they are social scientists might seem like special pleading when coming from a sociologist, but social autopsies, such as those suggested by Eric Klinenberg[5] in his examination of the deadly Chicago heat wave of 1995, are not frivolous, both after the fact and, as training, prior to the danger. In the case of the deadly 1990 Plainfield tornado, recognition of the need to be proactive could have saved lives. If meteorologists are public scientists whose goal is the protection of human life and property, it is essential that they conceive of their job as being more than reading the skies, but as reading the streets.

The weather service infuses our lives with science—or what we take to be science, despite the ways that it is shaped given social conditions. Local offices with their responsibility for the daily forecast and their requirement to warn for severe weather; the Environmental Modeling Center that produces model guidance, based on observations, equations, and theory that shape what will be forecast; and the Storm Prediction Center that primes and sensitizes the public to be prepared to act, each contributes to public science. We could not do without this work.

Any scene when examined intensively appears odd, and operational meteorology is no different. People do not always act as we imagine they should, but they act in ways similar to how *we* act in our social worlds. They share and create idiocultures, adjust to organizational constraints, and develop routine forms of action with those who share their spaces.

Weather matters, and so do those who reveal it. We could hardly imagine a modern society without the routine, confident, and shared claims of meteorology. Forecasters fight for this right and are challenged by nature's autonomy. Over time they are able to persuade, to justify, and to know. In this, they become authors of the storm.

Notes

PREFACE

1. The reality that much of the public is riveted by the weather and believes that they understand it is a mixed blessing for meteorologists, who find nonspecialists looking over their shoulder, sometimes without appreciating what they are seeing. As a result, to gain some measure of autonomy, meteorologists may create expertise through user-unfriendly jargon.

2. The defining, ur-texts in this tradition are Bruno Latour and Steven Woolgar, *Laboratory Life: The Construction of Scientific Facts* (Beverly Hills: Sage, 1979), and Karin Knorr-Cetina and Michael Mulkay, eds., *Science Observed: Perspectives on the Social Study of Science* (London: Sage, 1983).

3. David N. Livingstone, *Putting Science in Its Place: Geographies of Scientific Knowledge* (Chicago: University of Chicago Press, 2003).

4. Bruno Latour, *Science in Action* (Cambridge: Harvard University Press, 1987), pp. 232–47.

5. W. Henry Lambright and Stanley A. Changnon, Jr., "Arresting Technology: Government, Scientists, and Weather Modification," *Science, Technology and Human Values* 14 (1989): 340–59.

6. Congress has recently debated the extent to which the National Weather Service should be permitted to communicate with the public, essentially competing with private concerns. Of course, the private concerns get their data from the technology of the NWS. The debate is, thus, whether only private organizations should be able to access this government-gathered data.

7. Anthony Giddens, *Modernity and Self-Identity* (Stanford: Stanford University Press, 1991), pp. 109–43.

INTRODUCTION

1. Verlyn Klinkenborg, "The Moral Dimension of Weather in an Age of Forecasts from Everywhere," *New York Times,* November 6, 2003, p. A32.

2. Steve Matthewman, "Weather Modification in South Africa: Public Reactions to a 'Social Science,'" *Society-in-Transition* 29 (1998): 104–17; Barbara Farhar, "The Public Decides about Weather Modification," *Environment and Behavior* 9 (1977): 279–310.

3. The military also hires meteorologists, as does NASA and the FAA, although NWS meteorologists staff the FAA air traffic control centers.

4. Public science has been used with a slightly different meaning by historian Frank Turner, who speaks of public science in light of the desire of certain nineteenth-century British scientists to make their conclusions available to the general public. Here I refer to a science that in its essential character operates in the public domain with various publics as its primary audience. See Frank M. Turner, "Public Science in Britain, 1880–1919," *Isis* 71, no. 4 (1980): 589–608; Thomas F. Gieryn, *The Cultural Boundaries of Science: Credibility on the Line* (Chicago: University of Chicago Press, 1999), pp. 40–43.

5. The Massachusetts Institute of Technology has such a department, but, of course, MIT has a science and technology focus, unlike their Cambridge neighbor.

6. Philip Selznick, *TVA and the Grassroots* (Berkeley: University of California Press, 1949).

7. H. M. Collins and Robert Evans, "The Third Wave of Science Studies: Studies of Expertise and Experience," *Social Studies of Science* 32 (2002): 235–96.

8. The National Weather Service maintains a website that explains their formal acronyms (http://www.nwstc.noaa.gov/d.info/Acronym.HMTL).

9. Sharon Traweek, *Beamtimes and Lifetimes: The World of High Energy Physics* (Cambridge: Harvard University Press, 1988), p. 19.

10. See, famously, Malcolm Cowley, "Sociological Habit Patterns in Transmogrification," *Reporter* 20 (September 20, 1956): 44ff.

11. Jan Golinski, *Making Natural Knowledge: Constructivism and the History of Science* (Cambridge: Cambridge University Press, 1998), p. 122; Karin Knorr-Cetina, *Epistemic Cultures: How the Sciences Make Knowledge* (Cambridge: Harvard University Press, 1999), p. 112; Rik Scarce, *Fishy Business: Salmon, Biology, and the Social Construction of Nature* (Philadelphia: Temple University Press, 2000), p. 101.

12. Richard Hamblyn, *The Invention of Clouds* (New York: Farrar, Straus, & Giroux, 2001), pp. 149–50; John D. Cox, *Weather for Dummies: A Reference for the Rest of Us!* (Foster City, CA: IDG Books, 2000), p. 38.

13. Andrew Ross, "The Work of Nature in the Age of Electronic Emission," *Social Text* 18 (1987/88): 116–28, p. 123.

14. The National Weather Service is moving away from the clear status hierarchy associated with the distinction of "lead" and "journeymen" to the more subtle "senior" and "general" forecasters. At one point, before a forecast could be issued it had to be read by a lead forecaster, but today, whatever the official policy, this is rarely done. While offices should have one lead forecaster on duty at all times, because of vacations and other scheduling issues, this is not always the case.

15. Howard S. Becker, Blanche Geer, Everett C. Hughes, and Anselm L. Strauss, *Boys in White: Student Culture in Medical School* (Chicago: University of Chicago Press, 1961).

16. It seemed disingenuous to disguise the name of this office, because so much of the weather had to do with Chicago, but the names and sometimes the positions of individuals are shifted. The names of the two other Midwestern offices have been changed.

17. That the top meteorology programs are at Pennsylvania State University, Florida State University, and the University of Oklahoma reveals something about the status ranking of this science, although Michigan, Wisconsin, and MIT (in a more specialized way) have fine programs as well.

18. The SPC is now increasingly concerned with forecasting winter storms and conditions conducive to fire outbreaks.

19. The tracking of large systems is the crucial problem for meteorologists. Because of the long lead time, the problem of emergence is of less moment. In contrast, emergence is crucial for tornadoes and floods.

20. G. K. Grice, R. J. Trapp, S. F. Corfidi, R. Davies-Jones, C. C. Buonanno, J. P. Craven, K. K. Droegemeier, C. Duchon, J. V. Houghton, R. A. Prentice, G. Romine, K. Schlacter, and K. K. Wagner, "The Golden Anniversary Celebration of the First Tornado Forecast," *Bulletin of the American Meteorological Society* 80 (July 1999): 1341–48.

21. Stephen F. Corfidi, "The Birth and Early Years of the Storm Prediction Center," *Weather and Forecasting* 14 (1999): 507–25, p. 507.

22. I thank Christena Nippert-Eng for this insight.

23. John D. Cox, *Weather for Dummies.*

24. See Everett Hughes, *The Sociological Eye: Selected Papers on Work, Self, and the Study of Society* (Chicago: Aldine, 1971).

25. Only one forecaster informed me of concern about my presence. In the early stages of my research this man was supportive, but when I returned from a summer break, he informed me that he did not wish me to observe his actions, apparently because he noted me taking notes about matters that he considered embarrassing. Eventually, after I shared some previous writing, our positive relationship was renewed, but, unlike many others, there were some questions in the interview that he preferred not to answer.

26. Frederick K. Lutgens and Edward J. Tarbuck, *The Atmosphere: An Introduction to Meteorology* (8th ed.; Upper Saddle River, NJ: Prentice Hall, 2000), pp. 320–21.

27. Aristotle, *Meteorologica,* trans. H. D. P. Lee (London: W. Heinemann, 1962).

28. Klinkenborg, "The Moral Dimension of Weather."

29. Richard Hamlyn, *The Invention of Clouds* (New York: Farrar, Strauss & Giroux, 2001). Also see Jenny Uglow, *The Lunar Men: Five Friends Whose Curiosity Changed the World* (New York: Farrar Straus & Giroux, 2002).

30. Ibid., 75–93.

31. Ibid., 156.

32. The strong National Weather Service identity has proven problematic for NOAA leaders who desire a more visible corporate identity for themselves, which they are attempting to achieve by diminishing the prominence of the NWS brand. I was told by a headquarters officer that General Kelly, while at NWS, decreed that the NOAA logo replace the NWS logo in all PowerPoint presentations, and that the weather service be referred to publicly as "NOAA's National Weather Service." This attempt at public relations erasure apparently reached the point where t-shirts and polo shirts bearing the NWS logo are no longer sold in the NOAA store in Silver Spring. So far there is no indication that NOAA has become a household name.

33. Cox, *Weather for Dummies,* 16.

34. Diane Vaughan, *The Challenger Launch Decision* (Chicago: University of Chicago Press, 1996), p. 403.

35. Tornadoes are often known by the place that they occurred, a choice of the meteorologist as tornadoes may move through several locations. In the study I discuss the 1990 Plainfield (IL) tornado, the 1999 Oklahoma City tornadoes (although its most serious damage was in Moore, Oklahoma), and the 2002 La Plata (MD) tornado. Unlike hurricanes, they are not given a human face. Perhaps they are not around long enough for us to come to know them.

36. Keli Pirtle Tarp, "Communication in the Distributed Cognition Framework: An Ethnographic Study of a National Weather Service Office," manuscript, University of Oklahoma, 2001.

37. Leigh Thompson and Gary Alan Fine, "Socially Shared Cognition, Affect and Behavior: A Review and Integration," *Personality and Social Psychology Review* 3 (1999): 278–302.

38. Jeff Rosenfeld, "Looking Backward for the Future," *Bulletin of the American Meteorological Society* 84 (October 2003): 1328. On the general distinction between formal, codified, algorithmic knowledge and practical, hands-on, experiential knowledge, see Harry Collins, "The Seven Sexes: A Study in the Sociology of a Phenomenon, or the Replication of Experiments in Physics," *Sociology* 9 (1975): 205–24.

39. Mark Monmonier, *Air Apparent: How Meteorologists Learned to Map, Predict, and Dramatize Weather* (Chicago: University of Chicago Press, 1999), p. 10.

40. Ibid., p. 15.

41. Stephen Cole, "The Hierarchy of the Sciences?" *American Journal of Sociology* 89 (1983): 111–39.

42. Gary Alan Fine, "Small Groups and Cultural Creation: The Idioculture of Little League Baseball Teams," *American Sociological Review* 44 (1979): 733–45.

43. Steven Shapin, *A Social History of Truth: Civility and Science in Seventeenth-Century England* (Chicago: University of Chicago Press, 1994).

44. Eviatar Zerubavel, *The Fine Line: Making Distinctions in Everyday Life* (New York: Free Press, 1991).

CHAPTER ONE

1. The description in this section is a synthesis of events that I have witnessed in local meteorological offices. This analysis doesn't represent any one particular day, but provides the reader a *feel* for meteorological work.

2. Tim Vasquez, "Tornado!: An Evening with the Fort Worth Texas, National Weather Service," *Weatherwise* 56 (November–December 2003): 33–38, p. 36.

3. "Meteorology Makes the Top 10," *Bulletin of the American Meteorological Society* 83 (September 2002): 1302.

4. Gary Alan Fine, *Kitchens: The Culture of Restaurant Work* (Berkeley: University of California Press, 1996).

5. This siting emphasizes the importance of geography in the creation of scientific spaces. The location of radar equipment determines what is observed, when, and with what precision. See David N. Livingstone, *Putting Science in Its Place: Geographies of Scientific Knowledge* (Chicago: University of Chicago Press, 2003).

6. Personal communication, Shripad (Jayant) Deo, 2004.

7. Thomas Gieryn, "What Buildings Do," *Theory and Society* 31 (2002): 35–74.

8. Livingstone, *Putting Science in Its Place.*

9. More formally it is termed the operations area, but in informal talk it is often labeled the floor.

10. There are also four meteorologists employed at the Air Traffic Control Center in Aurora, but their responsibility is for weather that affects airplanes after take-off and before landing.

11. In some offices, the lights are bright and flourescent, keeping people awake; in other offices they are muted, providing a calming environment. Some offices are carpeted, some not, creating different levels of ambient noise, producing more or less distractions, fitting into and contributing to the office culture. The Chicago office was known as noisy and distracting, and hence more social. Daily distractions could be so insistent that some forecasters asserted that their midnight forecasts were more accurate than those in the afternoon, despite their exhaustion. Offices also differ in whether there is music in the background, and, if so, who gets to select—for some, a perk of status.

12. At one point the lead forecaster was required to be in charge of the public forecast, but this is no longer the case. When administrators, notably the MIC, fill in, however, they routinely work the less technologically advanced aviation desk.

13. Other offices divide the workload into long-term and short-term forecasts, with short-term forecasters having the responsibility for those forecasts that apply to the next few hours and the long-term forecaster having the responsibility for the "extended" forecast. That the extended forecast during the time of this research was for a week, not six or eight days, is an example of the

importance of the cognitive organization of time (see Eviatar Zerubavel, *The Seven Day Circle* [Chicago: University of Chicago Press, 1989]). Since the research ended, the extended forecast covers ten days, not nine or eleven.

14. Christena Nippert-Eng, *Home and Work* (Chicago: University of Chicago Press, 1996); Thomas Gieryn, "Biotechnology's Private Parts (and Some Public Ones)," in Crosbie Smith and Jon Agar, eds., *Making Space for Science* (London: Macmillan, 1998), pp. 281–312.

15. Nippert-Eng, *Home and Work,* pp. 145–48.

16. Eviatar Zerubavel, *The Fine Line* (New York: Free Press, 1991). Bus drivers apparently do something similar when they adjust the mirrors, set seat positions, and clean the dashboards, even placing pictures of their family within view. They, too, are just passing through.

17. Mary Douglas, *Purity and Danger* (London: Routledge & Kegan Paul, 1966).

18. At one point some faculty clubs, such as that at Harvard, had bars that remained open all afternoon. University faculty had the strategically problematic task of ordering a drink (or several) without appearing to be alcoholics, maintaining a nice academic role distance. Others had couches in their offices for amorous interludes.

19. Stephen Barley, "Technology as an Occasion for Structuring: Evidence from Observations of CT Scanners and the Social Order of Radiology Departments," *Administrative Science Quarterly* 31 (1986): 78–108.

20. Joan Fujimura, *Crafting Science: A Sociohistory of the Quest for the Genetics of Cancer* (Cambridge: Harvard University Press, 1996), p. 12.

21. Adele E. Clarke and Joan H. Fujimura, "Which Tools? Which Jobs? Why Right?" in Adele E. Clarke and Joan H. Fujimura, eds., *The Right Tools for the Job: At Work in Twentieth-Century Life Sciences* (Princeton: Princeton University Press, 1992), p. 9.

22. Howard Becker, *Art Worlds* (Berkeley: University of California Press, 1982), p. 1.

23. Clarke and Fujimura, "Which Tools? Which Jobs? Why Right?"

24. Michael Flaherty, *A Watched Pot: How We Experience Time* (New York: New York University Press, 1999).

25. Robert Merton, *The Sociology of Science* (Chicago: University of Chicago Press, 1973).

26. Warren O. Hagstrom, *The Scientific Community* (New York: Basic Books, 1965).

27. David M. Wegner, "Transactive Memory: A Contemporary Analysis of the Group Mind," in B. Mullen and G. R. Goethals, eds., *Theories of Group Behavior* (New York: Springer, 1987), pp. 185–208; E. T. Higgins, "The 'Communication Game': Implications for Social Cognition and Persuasion," in E. T. Higgins, C. P. Herman, and Mark P. Zanna, eds., *Social Cognition: The Ontario Symposium* (Hillsdale, NJ: Lawrence Erlbaum Associates, 1981), pp. 343–92; Leigh Thompson and Gary Alan Fine, "Socially Shared Cognition, Affect and Behavior: A Review and Integration," *Personality and Social Psychology Review* 3 (1999): 278–302; Edwin Hutchins, *Cognition in the Wild*

(Cambridge: MIT Press, 1995); Charles Goodwin, "Seeing in Depth," *Social Studies of Science* 25 (1995): 237–74.

28. Steven Shapin, "The Invisible Technician," *American Scientist* 77, no. 6 (1989): 554–63.

29. Chandra Mukerji notes in this vein: "The tradition of ignoring laboratory support personnel fits nicely with the view of machine tenders as deskilled. Lab assistants are easily conceptualized as robotlike followers of directives from their superiors and their machinery" (*A Fragile Power: Scientists and the State* [Princeton University Press, 1989], p. 136). See also Park Doing, "Lab Hands and the 'Scarlet O': Epistemic Politics and (Scientific) Labor," *Social Studies of Science* 34 (2004): 299–323; Daniel Chambliss, *Beyond Caring: Hospitals, Nurses, and the Social Organization of Ethics* (Chicago: University of Chicago Press, 1996).

30. All forecasters were male in Belvedere and Chicago during this research.

31. Deborah Tannen, *Talking from 9 to 5: Women and Men in the Workplace* (New York: Avon, 1995).

32. That this may not always be true in each office is suggested by a comment made by a journeyman forecaster who, visiting the Chicago office for training purposes, jokes, "I see it as my personal responsibility to give the leads as hard a time as I can" (Field notes). It would be hard to imagine this being said by someone in any of the three offices studied; the status divisions were carefully covered.

33. One informant, a NWS fellow, used the metaphor of the local offices as constituting "122 Navy ships" (Joe Banas, personal communication, 2004).

34. The term midnight covers not the moment itself, but the hours after the moment, and not the hours before, which are labeled the evening shift. Four a.m. cognitively falls within the domain of midnight in a way that 8:00 p.m. does not. Late night workers have their own culture and set of organizations. See Murray Melbin, *Night as Frontier: Colonizing the World after Dark* (New York: Free Press, 1987).

35. Marty Klein, *Shift Workers' Handbook* (Lincoln, NE: Syncho Tech, 1997), pp. 1–2. Even if some of the data should be viewed with caution, it indicates the rhetorical threats under which shift workers operate and feel themselves to operate.

36. For a related case in which some oceanographic scientists need to leave home for considerable periods of time, see Mukerji, *A Fragile Power,* 72.

37. Gary Alan Fine, "Organizational Time: Temporal Demands and the Experience of Work in Restaurant Kitchens," *Social Forces* 69 (1990): 95–114.

38. Eviatar Zerubavel, *Patterns of Time in Hospital Life* (Chicago: University of Chicago Press, 1979).

39. For a related treatment of time at work, see Leslie Perlow, *Finding Time: How Corporations, Individuals, and Families Can Benefit from New Work Practices* (Ithaca: ILR Press, 1997).

40. Roger Pielke, Jr., and R. E. Carbone, "Weather Impacts, Forecasts, and Policy," *Bulletin of the American Meteorological Society* 83 (March 2002): 396.

41. Daniel McCarthy, "Tornado Trends across the United States, 1950–2000," paper presented at the National Weather Association Annual Meeting, 2001.

42. Warnings are also issued for severe winter weather (heavy snow or blizzards), high winds, flooding, cold, and heat, but I focus on severe thunderstorm and tornado warnings.

43. In this way these organizations are like 'near-groups' (Lewis Yablonsky, *The Violent Gang* [New York: Macmillan, 1962]), quiescent, except when they are where the action is.

44. Personal communication, Shripad (Jayant) Deo, 2004.

45. The younger forecasters are reputedly more eager for severe weather. An older forecaster told me, "When you're younger, you kind of live for enjoying that, but not anymore. I hope for a season when there is no severe weather."

46. Gary Alan Fine, "Public Narration and Group Culture: Discerning Discourse in Social Movements," in Hank Johnston and Bert Klandermans, eds., *Social Movements and Culture* (Minneapolis: University of Minnesota Press, 1995), pp. 127–43.

47. Erving Goffman, *Interaction Ritual* (New York: Anchor, 1967), pp. 149ff.

48. For a sociological analysis of that event, from outside the meteorological perspective, see Eric Klinenberg, *Heat Wave* (Chicago: University of Chicago Press, 2002).

49. Gary England, an Oklahoma broadcast meteorologist, refers to the "surge of excitement" (Gary A. England, *Weathering the Storm* [Norman: University of Oklahoma Press, 1996], p. 201). See also David Laskin, *Braving the Elements: The Stormy History of American Weather* (New York: Anchor, 1996), p. 151.

50. In contrast, one meteorologist told me that "freezing rain scares me," not tornadoes.

51. Randall Collins, *Interaction Ritual Chains* (Princeton: Princeton University Press), 2004.

52. Rik Scarce, *Fishy Business: Salmon, Biology, and the Social Construction of Nature* (Philadelphia: Temple University Press, 2000), p. 148.

53. Robert Kohler, *Lords of the Flies: Drosophila Genetics and the Experimental Life* (Chicago: University of Chicago Press, 1994).

54. Mukerji, *A Fragile Power*, pp. 70–71.

55. Marianne Paget, *The Unity of Mistakes: A Phenomenological Interpretation of Medical Work* (Philadelphia: Temple University Press, 1988), p. 149.

56. G. Nigel Gilbert and Michael Mulkay, *Opening Pandora's Box: A Sociological Analysis of Scientists' Discourse* (Cambridge: Cambridge University Press, 1984), chap. 4.

57. Marvin B. Scott and Stanford M. Lyman, "Accounts," *American Sociological Review* 33 (1968): 46–62.

58. Charles Bosk, *Forgive and Remember: Managing Medical Failure* (Chicago: University of Chicago Press, 1979).

59. The extreme case is that of Captain Robert Fitzroy, head of the British Meteorological Department. After much criticism of his storm forecasts, he

committed suicide in 1865 in despair (Jack Fishman and Robert Kalish, *The Weather Revolution: Innovations and Imminent Breakthroughs in Accurate Forecasting* [New York, Plenum, 1994], p. 4).

60. John Seabrook, "The No-Show Snow," *New Yorker*, March 19, 2001, p. 49.

61. Arlie Russell Hochschild, *The Managed Heart: Commercialization of Human Feeling* (Berkeley: University of California Press, 1983).

62. Carol Z. Stearns and Peter N. Stearns, *Anger: The Struggle for Emotional Control in America's History* (Chicago: University of Chicago Press, 1989).

63. Randall Collins, *Interaction Ritual Chains.*

64. Females tend to have greater fear of severe weather than men; see Ronald A. Kleinknecht, "Afraid of the Weather?" *Weatherwise* 55 (November–December 2002): 15–16.

65. Fine, *Kitchens,* pp. 64–67.

66. Flaherty, *A Watched Pot: How We Experience Time.*

67. Erving Goffman, *Frame Analysis* (Cambridge: Harvard University Press, 1974).

68. Harold Garfinkel, Michael Lynch, and Eric Livingston, "The Work of Discovering Science Constructed with Material from the Optically Discovered Pulsar," *Philosophy of the Social Sciences* 11 (1981): 131–58.

69. This becomes particularly notable when one office passes off a storm to the office to the north or east, and there may be deference to the established ground truth, even as storms keep evolving. One forecaster reports: "[They] were passing off to us a storm with a history of tornadoes . . . even on that basis I would issue a warning, even if the radar data are not fully persuasive" (Field notes).

70. Tests are underway to move away from parallelograms to other geometric shapes that better capture storm activity.

71. As a result, Chicago believes of the SPC that "they usually overblow everything" (Field notes), and, in turn, SPC feels that when they call Chicago to coordinate, they will often get frank discussions or complaints (depending who describes the conversation).

72. Carol Heimer, *Reactive Risk and Rational Action: Managing Moral Hazard in Insurance Contracts* (Berkeley: University of California Press, 1985); Lee Clarke, *Acceptable Risk? Managing Decisions in a Toxic Environment* (Berkeley: University of California Press, 1989); Gary Alan Fine and Lori Holyfield, "Secrecy, Trust, and Dangerous Leisure: Generating Group Cohesion in Voluntary Organizations," *Social Psychology Quarterly* 59 (1996): 22–38; Mary Douglas and Aaron Wildavsky, *Risk and Culture: An Essay on the Selection of Technical and Environmental Dangers* (Berkeley: University of California Press, 1982).

73. James F. Short, Jr., "The Social Fabric at Risk: Toward the Social Transformation of Risk Analysis," *American Sociological Review* 49 (1984): 711–25.

74. Diane Vaughan, *The Challenger Launch Decision* (Chicago: University of Chicago Press, 1996), p. 400.

75. Amos Tversky and Daniel Kahneman, "Judgment under Uncertainty: Heuristics and Biases," *Science* 185 (1974): 1124–31.

CHAPTER TWO

1. Thomas F. Gieryn, *Cultural Boundaries of Science: Credibility on the Line* (Chicago: University of Chicago, 1999), pp. 1–35.

2. Chandra Mukerji, *A Fragile Power: Scientists and the State* (Princeton: Princeton University Press, 1989), p. ix.

3. Joseph C. Hermanowicz, *The Stars Are Not Enough: Scientists—Their Passions and Professions* (Chicago: University of Chicago Press, 1998).

4. Robert E. Kohler, *Landscapes and Labscapes: Exploring the Lab-Field Border in Biology* (Chicago: University of Chicago Press, 2002), p. 98.

5. Jack Fishman and Robert Kalish (*The Weather Revolution: Innovations and Imminent Breakthroughs in Accurate Forecasting* [New York: Plenum, 1994], p. 29) suggest that "Meteorologists . . . are the Rodney Dangerfields of science. They get no respect from the formal scientific disciplines such as physics and chemistry."

6. Pierre Bourdieu, "The Peculiar History of Scientific Reason," *Sociological Forum* 6 (1991): 3–26.

7. Bruno Latour and Steve Woolgar, *Laboratory Life: The Construction of Scientific Facts* (Beverly Hills: Sage, 1979); Karin Knorr-Cetina, *The Manufacture of Knowledge: An Essay on the Constructionist and Contextual Nature of Science* (Oxford: Pergamon, 1981); Sharon Traweek, *Beamtimes and Lifetimes: The World of High-Energy Physicists* (Cambridge: Harvard University Press, 1988).

8. Diane Vaughan, "The Role of the Organization in the Production of Techno-Scientific Knowledge," *Social Studies of Science* 29 (1999): 913–43.

9. Leslie Sklair, *Organized Knowledge: A Sociological View of Science and Technology* (St. Albans: Hart-Davis MacGibbon, 1973), p. 69. For Merton's discussion, see Robert Merton, "Science and Democratic Social Structure," in Robert K. Merton, *Social Theory and Social Structure* (New York: Free Press, 1968), pp. 604–15.

10. An indication of the status associated with research occurred at the Storm Prediction Center, where forecasters do publish some descriptive analyses. One forecaster who had previously published such an article was sent a manuscript to review by that journal. He made it clear how seriously he was taking the project, attempting to find a series of days to get into the analysis, seeing the request as an honor, rather than as the burden that many of his more scientific colleagues would have felt it to be.

11. Lynn Eden, *Whole World on Fire: Organizations, Knowledge and Nuclear Devastation* (Ithaca: Cornell University Press, 2003).

12. What research there is often appears on the local website or in internal agency communiques, rather than being published in refereed journals. This analysis does not apply to the research centers of the National Weather Service, such as the Environmental Modeling Center in Camp Springs, Maryland, or the

Storm Prediction Center in Norman, Oklahoma, which shares its space with the more explicitly research-oriented National Severe Storm Laboratory.

13. Karin Knorr-Cetina, "New Developments in Science Studies: The Ethnographic Challenge," *Canadian Journal of Sociology* 8 (1983): 153–77.

14. Michael Lynch, *Scientific Practice and Ordinary Action: Ethnomethodology and Social Studies of Science* (Cambridge: Cambridge University Press, 1985), p. 315.

15. David Laskin, *Braving the Elements: The Stormy History of American Weather* (New York: Anchor, 1996), p. 189.

16. Gary Alan Fine, "Justifying Work: Occupational Rhetorics as Resources in Restaurant Kitchens," *Administrative Science Quarterly* 41 (1996): 90–115.

17. John Law and Michael Lynch, "Lists, Field Guides and the Descriptive Organization of Seeing: Birdwatching as an Exemplary Observational Activity," *Human Studies* 11 (1988): 271–303.

18. Warren O. Hagstrom, *The Scientific Community* (New York: Basic Books, 1965), pp. 1, 295.

19. At both the National Center for Environmental Prediction and at the Storm Prediction Center and National Severe Storm Laboratory, the linkage between prediction and scientific inquiry are combined in the institution of a half-hour discussion each weekday (the academics rarely work on weekends). As one said about this meeting, it brings "forecast and research together" (Field notes). But the questions reveal the different interests and contexts of knowledge. The discussions have a different feel depending on whether a forecaster or a researcher is leading the discussion, and one forecaster told me that at times he couldn't understand the discussions when led by his colleagues from the research branch. One forecaster described these meetings from his vantage point: "For some people there is some apprehensiveness in getting up in front of a bunch of scientists. It can be unnerving. Scientists don't like sitting down at a work station and saying what they think will happen instead of looking back on things. Scientists are looking for precision. Those folks can be a bit of a pain. I can speak the language, but people can be on pins and needles. There is a culture clash" (Field notes). Of particular note is the temporal difference: forecasters look forward, researchers backwards.

There has been much less connection between these Oklahoma offices and the offices of the Department of Meteorology at the University of Oklahoma a few miles away, and, for that matter, little contact with the local forecast office across the street, putting out routine daily forecasts. On the drawing board is a plan to bring all of the meteorological offices together in a large new building on campus. Sometimes even a staircase, however, can pose an insuperable barrier.

20. Cultural studies scholar Andrew Ross ("The Work of Nature in the Age of Electronic Emission," *Social Text* 18 [1987/88]: 116–28, p. 119) seems to agree, noting "the 'gay science' of weather folklore and divination has been replaced by our new 'unhappy' relation to the legitimate scienticity of the meteorologist's enlightened jargon-laden language of forecasting; the everyday cult of 'experience' has been replaced by the professionalist ethos of 'expertise' in responding to weather semiology."

21. Despite this claim, however, as Nicholas Christakis (*Death Foretold: Prophecy and Prognosis in Medical Care* [Chicago: University of Chicago Press, 1999]) points out, doctors often shy away from prognosis, both because of the emotional strains and the uncertainty of prediction.

22. Jan Golinski, *Making Natural Knowledge: Constructivism and the History of Science* (Cambridge: Cambridge University Press, 1998), p. 98; Robert E. Kohler, *Landscapes and Labscapes: Exploring the Lab-Field Border in Biology* (Chicago: University of Chicago Press, 2002).

23. James Rodger Fleming, *Meteorology in America, 1800–1870* (Baltimore: Johns Hopkins University Press, 1990), p. xx.

24. H. M. Collins, *Changing Order: Replication and Induction in Scientific Practice* (Chicago: University of Chicago Press, 1985), p. 19.

25. Joel C. Curtis, "The Forecast Process—One Forecaster's Perspective," papers of the First Conference on Artificial Intelligence, January 11–16, 1998, pp. 151–56.

26. Joan Fujimura, *Crafting Science: A Sociohistory of the Quest for the Genetics of Cancer* (Cambridge: Harvard University Press, 1996).

27. The meteorologists I observed did not talk much about global warming. They were dubious about these long-term changes, felt they weren't experts, and believed that the National Oceanic and Atmospheric Administration wished them to speak with a single voice, and so they did not participate in political debates about climate change.

28. It is partially because operational meteorologists focus on the short term—often seeing unusual heat and cold—that many are suspicious of the climatologist's emphasis on "global warming." Whenever they, like the public, face a long hot spell, they are more willing, humorously, to accept global warming, as one remarked on a warm December afternoon: "This is global warming. I am now convinced of it. You can't argue with observation. When was the last time you saw any lows" (Field notes). In January after a period with little snow, another forecaster proclaimed, "I began to be a believer in global warming" (Field notes). More seriously many meteorologists emphasize that temperatures "naturally" and cyclically move up and down, and if it is warmer now, it will soon become colder than normal. One meteorologist even penned a poem entitled "Chilly Thoughts on Global Warming" (Field notes).

29. It is symbolic that the offices of the AMS are located in Boston's tony Beacon Hill, while, in 2001, the offices of the NWS were located in Montgomery, Alabama.

30. The collaboration is real, and employees of the two organizations sometimes collaborate on papers (see John S. Kain, Paul R. Janish, Steven J. Weiss, Michael E. Baldwin, Russell S. Schneider, and Harold E. Brooks, "Collaboration between Forecasters and Research Scientists at the NSSL and the SPC: The Spring Program," *Bulletin of the American Meteorological Society* 84 [December 2003]: 1797–1806). Yet, I found it striking that the keys of the staff of NSSL could not open the door of the SPC forecast office.

31. Karin Knorr-Cetina, *The Manufacture of Knowledge: An Essay on the Constructivist and Contextual Nature of Science* (Oxford: Pergamon, 1981); Jason Owen-Smith, "Managing Laboratory Work through Skepticism:

Processes of Evaluation and Control," *American Sociological Review* 66 (2001): 427–52.

32. Golinski, *Making Natural Knowledge,* p. 167

33. Clifford Geertz, *Local Knowledge: Further Essays in Interpretive Anthropology* (New York: Basic Books, 1983), p. 157.

34. Gary Alan Fine, "Small Groups and Cultural Creation: The Idioculture of Little League Baseball Teams," *American Sociological Review* 44 (1979): 733–45.

35. Nina Eliasoph and Paul Lichterman, "Culture in Interaction," *American Journal of Sociology* 108 (2003): 735–94; Tom McFeat, *Small-Group Cultures* (New York: Pergamon, 1974).

36. Gary Alan Fine, *With the Boys: Little League Baseball and Preadolescent Culture* (Chicago: University of Chicago Press, 1987), p. 125.

37. Andrew Pickering, "The Mangle of Practice: Agency and Emergence in the Sociology of Science," *American Journal of Sociology* 99 (1993): 559–89.

38. August Hollingshead, "Behavior Systems as a Field for Research," *American Sociological Review* 4 (1939): 816–22, p. 816; Alan Dundes, "Who Are the Folk?" in William Bascom, ed., *Frontiers of Folklore* (Boulder: Westview, 1977).

39. Herbert Blumer, *Symbolic Interactionism* (Englewood Cliffs: Prentice-Hall, 1969); Eliasoph and Lichterman, "Culture in Interaction."

40. Randy Hodson, *Working with Dignity* (Cambridge: Cambridge University Press, 2001).

41. Michael Mulkay and G. Nigel Gilbert, "Joking Apart: Some Recommendations Concerning the Analysis of Scientific Culture," *Social Studies of Science* 12 (1982): 585–613, p. 585; Mukerji, *A Fragile Power,* pp. 71–72.

42. Karin Knorr-Cetina, *Epistemic Cultures: How the Sciences Make Knowledge* (Cambridge: Harvard University Press, 1999), p. 203.

43. Ibid., p. 106; Diane Vaughan, *The Challenger Launch Decision: Risky Technology, Culture, and Deviance at NASA* (Chicago: University of Chicago Press, 1996), chap. 2.

44. Ibid.; Sharon Traweek, *Beamtimes and Lifetimes: The World of High Energy Physics* (Cambridge: Harvard University Press, 1988).

45. Traweek, *Beamtimes and Lifetimes,* p. 74.

46. Erving Goffman, *Forms of Talk* (Philadelphia: University of Pennsylvania Press, 1981), p. 46.

47. Ludwik Fleck, *Genesis and Development of a Scientific Fact* (Chicago: University of Chicago Press, 1979).

48. See, for example, Sharon Traweek, "Border Crossings: Narrative Strategies in Science Studies and among Physicists in Tsukuba Science City, Japan," in Andrew Pickering, ed., *Science as Practice and Culture* (Chicago: University of Chicago Press, 1992), pp. 429–65.

49. H. M. Collins, "The Meaning of Data: Open and Closed Evidential Cultures in the Search for Gravitational Waves," *American Journal of Sociology* 104 (1998): 293–338.

50. Peter Gallison, *Image and Logic: A Material Culture of Microphysics* (Chicago: University of Chicago Press, 1997), pp. 781–844.

51. Warren O. Hagstrom, *The Scientific Community* (New York: Basic Books, 1965), p. 293.

52. One administrator referred to his forecasters as "salmon." By this he meant that when they became lead forecasters, they tend to return to the community where they were raised—to spawn or to die. Perhaps there was a desire to hire "locals," whose understanding of the local meteorological conditions was bred in the bone. Because of the belief in the virtue of local knowledge, one forecaster told me that in his view at least half the forecasters on staff should be locals, because "they have an intuitive sense of the area" (Field notes).

53. This seemed particularly true at the Storm Prediction Center, where many forecasters have a deep fascination with severe weather and dream of being "where the action is."

54. Connections can arise from transferring from office to office, creating a subculture based on linked small groups (see Gary Alan Fine and Sherryl Kleinman, "Rethinking Subculture," *American Journal of Sociology* 85 [1979]: 1–20). Connections also emerge, however, from individuals in adjoining offices who share the same shift schedule and become friends through continuing coordination and through meetings and classes. The decrease in the number of gatherings has been detrimental to the establishment of a networked organization, creating instead what one administrator described as a "fragmented organization." At one point all forecast interns had to attend a three-week seminar where they met other members of their cohort, but that gathering has been phased out.

55. This is a point made by students of organizational culture, e.g. Gary Alan Fine, "Negotiated Orders and Organizational Cultures," *Annual Review of Sociology* 10 (1984): 239–62; John Weeks, *Unpopular Culture: The Ritual of Complaint in a British Bank* (Chicago: University of Chicago Press, 2004). There is still a need for comparative ethnographic analyses of organizations.

56. Fujimura, *Crafting Science*, p. 5; Karin Knorr-Cetina, *The Manufacture of Knowledge: An Essay on the Constructivist and Contextual Nature of Science* (Oxford: Pergamon Press, 1981).

57. Mukerji, *A Fragile Power*, p. ix.

58. Other offices answer the phone, "National Weather Service" or "National Weather Service, X office."

59. As weather conditions are largely unaffected by population, headquarters feels that staffing should be unaffected as well, despite the greater public safety concerns and demands from media, industry, and local government. Given the assumption that weather is weather, and populations are interchangeable, the weather service often does not consider the specific needs of their communities.

60. The MIC tried, upon arriving, to institute a dress code, but this was resisted by the staff, becoming a union issue, and eventually he retreated. He describes the situation: "I attempted to change the dress code at the office shortly after my arrival. I saw a lead forecaster interviewed on television with a tee shirt and blue jeans on and was appalled. . . . I coordinated with the regional director and posted the 'dress code.' . . . Once in effect, several individuals took offense and the union supported their stance. . . . the Union took it . . . to Weather Service Headquarters. [The director was too involved in NWS restructuring] and had

his hands full. So his response was to side with the union. . . . This was a hard hit. The 'time to strike' if you want to implement changes at an office is just after you arrive. I felt that if I could implement this 'code' it would eventually manifest a change in attitude and [instill] a positive 'esprit de corps.' Instead I started out with not only a defeat, but actually a step back." As a result of this symbolic defeat, in which the MIC was not supported by headquarters, any change in culture became problematic. Dress at the Chicago office, while not sloppy, was looser than at other offices, where jeans are not worn on day shifts and button-down shirts are.

61. Each office has something like this, especially useful for midnight shifts, but the Chicago office's was by far the most extensive.

62. Accountability operates in a dual fashion—as a means of bolstering individual autonomy and as a means of enforcing organizational control. Often these two modes of accountability conflict. Unless an agreement develops on the usage of legitimate authority, the creation of these trip wires becomes problematic.

63. Timothy P. Hallett, "Symbolic Power and the Social Organization of Turmoil: Order, Disruption, and Conflict in an Urban Elementary School," Ph.D. diss., Northwestern University, 2003.

64. Given the complaints that would likely arise in revision of the indulgency pattern, this new leadership would need to be supported by the higher levels of the organization; otherwise the opportunities for change would be blunted.

65. One Chicago forecaster described the well-liked MIC at Belvedere as a "little dictator" and described the Flowerland office as being in a "lock-down" (Interview).

66. Edward Rose and William Felton, "Experimental Histories of Culture," *American Sociological Review* 20 (1955): 383–92.

67. The office was not reminiscent of a fraternity party, and most of the time this office, like the others at which I observed, was quiet and serious. But joking and playful talk seemed more characteristic of the Chicago office, making it a friendlier and more comfortable place to observe.

68. Erving Goffman, *Frame Analysis* (Cambridge: Harvard University Press, 1974).

69. Richard G. Mitchell, *Mountain Experience: The Psychology and Sociology of Adventure* (Chicago: University of Chicago Press, 1983); Gary Alan Fine, *Morel Tales: The Culture of Mushrooming* (Cambridge: Harvard University Press, 1998), chap. 4.

70. Donald Roy, "Banana Time: Job Satisfaction and Informal Interaction," *Human Organization* 18 (1959/60): 156–68.

71. Susan Leigh Star and James R. Griesemer, "Institutional Ecology, 'Translations,' and Boundary Objects: Amateurs and Professionals in Berkeley's Museum of Vertebrate Zoology, 1907–39," *Social Studies of Science* 19 (1989): 387–420; also Geoffrey C. Bowker and Susan Leigh Star, *Sorting Things Out: Classification and Its Consequences* (Cambridge: MIT Press, 1999), p. 16.

72. Gary Alan Fine and Michaela DeSoucey, "Joking Cultures: Humor Themes as Social Regulation in Group Life," *Humor* 18 (2005): 1–22, pp. 2–4.

73. Richard E. DuWors, "Persistence and Change in Local Values of Two New England Communities," *Rural Sociology* 17 (1952): 207–17.

74. Fine, "Small Groups and Cultural Creation," p. 736.

75. J. L. Fischer, "Microethnology: Small-Scale Comparative Studies," in J. A. Clifton, ed., *Introduction to Cultural Anthropology* (Boston: Houghton Mifflin, 1967), pp. 75–87.

76. William B. Rogers and R. E. Gardner, "Linked Changes in Values and Behavior in the Out Island Bahamas," *American Anthropologist* 71 (1969): 21–35; Evan Z. Vogt and Thomas O'Dea, "Cultural Differences in Two Ecologically Similar Communities," *American Sociological Review* 18 (1953): 645–54.

77. A comparative analysis of office cultures, while desirable, is beyond the scope of this analysis. In a small *n* comparison, so many factors weigh upon the outcome that attempting to determine a single cause is apt to be misleading.

78. A similar pattern, although not quite as dramatic, was evident between Belvedere and one of their neighboring spin-up offices. As one of the administrators said of them, "They like to go for the home run, but they strike out a lot. They swing for the fences." He indicates that they often forecast heavy snow that doesn't usually materialize, but for which they get considerable media attention. I was told that the spin-up offices desire to demonstrate their independence from the older offices by these forecasts, transforming the forecasting process into a means of social differentiation.

79. One Chicago staff member wrote in response to an earlier version of this manuscript, "I don't think there is a hesitation to forecast or warn in dangerous events. . . . In an office which services ten million people any action is magnified by the number of people listening and (especially) the media. One must always be aware of the implied consequences or perhaps more the scale of the consequences when getting into a serious situation. More rural environments don't need to worry about reactions from cows or chickens!" This forecaster made the point that they were not derelict in their duty, but rather that the considerations in producing their forecasts would be different because of their social environment. Their conservatism was a function of the community that could be affected. However, because of the larger population at risk, one might argue that more warnings were called for.

80. This Flowerland forecaster told me that he used to write Area Forecast Discussions that were a page and a half long, sharing all of his meteorological knowledge, "I started on a planetary scale. That really annoyed Chicago. They said why don't you start with a galactic scale" (Field notes).

81. Diane Vaughan, *The Challenger Launch Decision* (Chicago: University of Chicago Press, 1996); James F. Short, Jr. "The Social Fabric at Risk: Toward the Social Transformation of Risk Analysis," *American Sociological Review* 49 (1984): 711–25.

82. Plainfield stands like a bookend with an event nine years later, a major tornado outbreak on May 3, 1999, in Oklahoma City, in which the local office issued warnings early and effectively. Of course, Oklahoma City was in the heart of tornado alley and early May was the heart of tornado season. As one forecaster at the SPC explained, "I think Oklahoma City in 1999 will be recalled

as a success of the weather service, just as Plainfield is recalled as the ultimate weather service disaster" (Field notes).

83. Julia Keller, "Utica, Illinois: A Wicked Wind Takes Aim," *Chicago Tribune,* December 5, 2004, pp. 1, 12–13. The fact that the office had a new MIC in charge less than two years allowed the success to be attributed to a change in office culture.

84. Most tornadoes in Illinois occur in the months of April to June.

85. Data about the storm is taken from the National Disaster Survey Report, *The Plainfield/Crest Hill Tornado* (Silver Spring, Maryland: National Weather Service, 1991).

86. Ibid., p. x.

87. Roberta Klein and Roger A. Pielke, Jr., "Bad Weather? Then Sue the Weatherman! Part I: Legal Liability for Public Sector Forecasts," *Bulletin of the American Meteorological Society* 83 (December 2002): 1791–99, p. 1795.

88. Vaughan, *The Challenger Launch Decision.*

89. One could blame God, but, unlike technology, the deity currently lacks agency within scientific and social scientific writing. He is no longer a legitimate social actor, edged off the stage of human events. Such a claim, normative once, would likely discredit the proposer.

90. Part of his poem, "WSFO Chicago Sings the Blues," reads:

> Our heads hang in shame
> For not playing the game
> To look low and high
> That we might verify
>
> Our repute has been had.
> Our press is all bad.
> From local to Region,
> From HQ and Cleveland.
>
> We're not taking this down!
> We'll, like, turn it around!
> Front Page Heroes we'll be!
> In due time, they'll ALL see!
>
> Let's not grovel or bark
> And curse at the dark.
> Light a candle within
> And watch that damned wind!

If the poem is unlikely to be anthologized, it is heartfelt, and ultimately hopeful.

91. I was told with some bitterness that the National Weather Service provided no counseling to employees, some of whom felt responsible for dozens of deaths. My informant pointed out that this stood in stark contrast to the aftermath of airline crashes or school shootings. He says, "No one asked if you were OK. No, I wasn't OK. It weighed on me heavily." Added to this was the fear that

individual forecasters might be sued. He concludes, "I can honestly say that if I knew all that, I wouldn't have started working for the weather service" (Field notes).

92. Michael Schudson, "How Culture Works: Perspectives from Media Studies on the Efficacy of Symbols," *Theory and Society* 18 (1989): 153–80.

93. Gary Alan Fine and Lori Holyfield, "Secrecy, Trust, and Dangerous Leisure: Generating Group Cohesion in Voluntary Organizations," *Social Psychology Quarterly* 59 (1996): 22–38.

CHAPTER THREE

1. Frederick K. Lutgens and Edward J. Tarbuck, *The Atmosphere: An Introduction to Meteorology* (eighth ed.; Upper Saddle River, NJ: Prentice Hall, 2000), pp. 320–21.

2. About 40 percent of all trained meteorologists work in the private sector (personal communication, Shripad Deo, 2004). Unlike government forecasters, these forecasters are not shielded from liability, even if much of their data derives from the NWS.

3. But see Wendell Bell, *The Foundation of Future Studies* (New Brunswick, NJ: Transaction, 1997); Fred Polak, *The Sociology of the Future* (New York: Oceania, 1961); Susan A. Van't Klooster and Marjolein B. A. Van Asselt. "Practicing the Scenario-Axes Technique," *Futures* 37 (2005): 1–16.

4. Tamotsu Shibutani, *Improvised News* (Indianapolis: Bobbs-Merrill, 1966).

5. Donald Levine, *Flight from Ambiguity* (Chicago: University of Chicago Press, 1988); Ulrich Beck, *Risk Society: Toward a New Modernity* (Newbury Park, CA: Sage, 1992).

6. Gordon Allport and Leo Postman, *The Psychology of Rumor* (New York: Holt, 1947).

7. Michael Katovich, "Identity, Time, and Situated Activity: An Interactionist Analysis of Dyadic Transactions," *Symbolic Interaction* 15 (1987): 25–47.

8. Richard L. Henshel, *On the Future of Social Prediction* (Indianapolis: Bobbs-Merrill, 1976); Richard L. Henshel, "Sociology and Social Forecasting," *Annual Review of Sociology* 8 (1982): 57–79.

9. Nicholas Rescher, *Predicting the Future: An Introduction to the Theory of Forecasting* (Albany: State University of New York Press, 1998).

10. Gideon Sjoberg, Elizabeth A. Gill, and Leonard D. Cain, "Countersystem Analysis and the Construction of Alternative Futures," *Sociological Theory* 21 (2003): 210–35, p. 215.

11. Paul N. Edwards, "Global Climate Science, Uncertainty and Politics: Data-Laden Models, Model-Filtered Data," *Science as Culture* 8 (1999): 437–72.

12. Howard S. Becker, "Art Worlds and Social Types," *American Behavioral Scientist* 19 (1976): 703–18.

13. Steven Shapin, *A Social History of Truth: Civility and Science in Seventeenth-Century England* (Chicago: University of Chicago Press, 1994).

14. Futurist research often addresses the first three points, but ignores the implications of the fourth. See, for example, Mario Bunge, "The Role of Forecast in Planning," *Theory and Decision* 3 (1973): 207–21, p. 207; Nicholas Rescher, *Predicting the Future: An Introduction to the Theory of Forecasting* (Albany: State University of New York Press, 1998), chap. 4.

15. The existence of legitimations of forecasting ability contributes to disciplinary status, a challenge that some see for sociology itself (Henshel, *On the Future of Social Prediction,* 41).

16. Peter T. Manicas, "Explaining the Past and Predicting the Future," *American Behavioral Scientist* 42 (1998): 398–405, p. 400; Nicholas Rescher, *Predicting the Future.*

17. Sjoberg, Gill, and Cain, "Countersystem Analysis and the Construction of Alternative Futures," p. 210.

18. Because of its ubiquity and because of the explicit usage of data and models within the context of an organizationally mandated big science, meteorology has been taken as a defining case of the process of forecasting. See Peter Wiles, "The Necessity and Impossibility of Political Economy," *History and Theory* 11 (1982): 3–14, p. 10; Theodore Caplow, "Imperfect Art or Imperfect Science?" *Tocqueville Review* 16 (1995): 119–33, p. 120.

19. Fred Davis, *Passage through Crisis: Polio Victims and Their Families* (New Brunswick, NJ: Transaction, 1991), p. 35; Nicholas Christakis, *Death Foretold: Prophecy and Prognosis in Medical Care* (Chicago: University of Chicago Press, 1999); Raphael Sassower and Michael Grodin, "Scientific Uncertainty and Medical Responsibility," *Theoretical Medicine* 8 (1987): 221–34.

20. Fred Davis ("Uncertainty in Medical Prognosis: Clinical and Functional," *American Journal of Sociology* 66 [1960]: 41–47) makes the point that sometimes medical practitioners—in his case those dealing with families who have children with polio—pretend to be uncertain about negative outcomes, leaving their clients time to reach sad realizations.

21. Eviatar Zerubavel, *Time Maps: Collective Memory and the Social Shape of the Past* (Chicago: University of Chicago Press, 2003).

22. This length is both a political and scientific matter. As the skills and technology of the operational forecaster improve, the length of time that meteorologists work with expands, altering the boundary with climatology (for a general analysis of occupational boundaries, see Andrew Abbott, *The System of Professions: An Essay on the Division of Expert Labor* [Chicago: University of Chicago Press, 1988]). In turn, the climatologist is able to extend knowledge on the other end, claiming a longer vision than before.

23. Levine, Flight from Ambiguity.

24. Trevor J. Pinch, "The Sun-Set: The Presentation of Certainty in Scientific Life," *Social Studies of Science* 11 (1981): 131–58, pp. 131–32, 154–55.

25. Simon Shackley and Brian Wynne, "Representing Uncertainty in Global Climate Change Science and Policy: Boundary-Ordering Devices and Authority," *Science, Technology and Human Values* 21 (1996): 275–302, p. 286.

26. Van't Klooster and Van Asselt, "Practicing the Scenario-Axes Technique."

27. Donileen Loseke, *Thinking about Social Problems* (Hawthorne, NY: Aldine, 1999).

28. Sara Delamont and Paul Atkinson, "Doctoring Uncertainty: Mastering Craft Knowledge," *Social Studies of Science* 31 (2001): 87–107.

29. Rik Scarce, *Fishy Business: Salmon, Biology, and the Social Construction of Nature* (Philadelphia: Temple University Press, 2000), p. 130.

30. Davis, "Uncertainty in Medical Prognosis," pp. 41–47; Marjolein B. A. van Asselt and Jessica Mesman, "Walking a Tightrope: A Comparative Analysis of Strategies for Dealing with Prognostic Uncertainty in Medical and Governmental Practices," paper presented at the Society for the Study of the Sociology of Science, Atlanta, GA, 2003.

31. Christakis, *Death Foretold*, p. 15.

32. Jeff Rosenfeld, "An Intuition for Chaos," *Bulletin of the American Meteorological Society* 84 (July 2003): 872.

33. Charles Goodwin, "Seeing in Depth," *Social Studies of Science* 25 (1995): 237–74.

34. Peter Gallison, *Image and Logic: A Material Culture of Microphysics* (Chicago: University of Chicago Press, 1997); Adele Clarke and Joan Fujimura, "Introduction," in Adele Clarke and Joan Fujimura, eds., *The Right Tools for the Job: At Work in Twentieth-Century Life Sciences* (Princeton: Princeton University Press, 1992), pp. 3–44.

35. Renee Ansbach, "Prognostic Conflict in Life-and-Death Decisions: The Organization as an Ecology of Knowledge," *Journal of Health and Social Behavior* 28 (1987): 215–31, p. 215.

36. Bruno Latour, *Science in Action: How to Follow Scientists and Engineers through Society* (Cambridge: Harvard University Press, 1987); Michel Callon, "Some Elements of a Sociology of Translation: Domesticating of the Scallops and the Fishermen of St. Brieuc Bay," in John Law, ed., *Power, Action and Belief: A New Sociology of Knowledge?* (London: Routledge & Kegan Paul, 1986), pp. 196–233.

37. Ivan Tchalakov, "The Object and the Other in Holographic Research: Approaching Passivity and Responsibility of Human Actors," *Science, Technology and Human Values* 29 (2004): 64–87, p. 64.

38. Mark Monmonier, *Air Apparent: How Meteorologists Learned to Map, Predict, and Dramatize Weather* (Chicago: University of Chicago Press, 1999), p. 146; see Ian Hacking, *Representing and Intervening: Introductory Topics in the Philosophy of Natural Science* (Cambridge: Cambridge University Press, 1983), p. 230; Jan Golinski, *Making Natural Knowledge: Constructivism and the History of Science* (Cambridge: Cambridge University Press, 1988), p. 32.

39. Barry Barnes, David Bloor, and John Henry, *Scientific Knowledge: A Sociological Analysis* (Chicago: University of Chicago Press, 1996), pp. 1–2.

40. Karin Knorr-Cetina, *The Manufacture of Knowledge: An Essay on the Constructionist and Contextual Nature of Science* (Oxford: Pergamon, 1981), p. 27.

41. Michael E. Lynch, "Sacrifice and the Transformation of the Animal Body into a Scientific Object: Laboratory Culture and Ritual Practice in the Neurosciences," *Social Studies of Science* 18 (1988): 265–89.

42. Chandra Mukerji, *A Fragile Power: Scientists and the State* (Princeton University Press, 1989), p. 106.

43. Knorr-Cetina, *The Manufacture of Knowledge,* pp. 116–17.

44. Delamont and Atkinson, "Doctoring Uncertainty."

45. Scientists often erase the importance of machine tenders, assuming their proper operation. The work in keeping these machines working is seen as external to office life. For the role of machine handler in the world of labor and the work of science, see David Halle, *America's Working Man* (Chicago: University of Chicago Press, 1984), pp. 127–30; Mukerji, *A Fragile Power*, p. 127; Park Doing, "Lab Hands and the 'Scarlet O': Epistemic Politics and (Scientific) Labor," *Social Studies of Science* 34 (2004): 299–323.

46. John Weeks, *Unpopular Culture: The Ritual of Complaint in a British Bank* (Chicago: University of Chicago Press, 2004).

47. Knorr-Cetina, The Manufacture of Knowledge, p. 122.

48. Jon Hindmarsh and Christian Heath, "Sharing the Tools of the Trade: The Interactional Constitution of Workplace Objects," *Journal of Contemporary Ethnography* 29 (2000): 523–62, p. 555.

49. In Flowerland, I was told about one small city, far from the radar, for which they often overforecast the amount of precipitation because the beam was aimed too high for an optimal reading (Field notes).

50. Jeff Rosenfeld, "Data Are Not Data," *Bulletin of the American Meteorological Society* 83 (June 2002): 836.

51. Traweek, "Border Crossings: Narrative Strategies in Science Studies and among Physicists in Tsukuba Science City, Japan," in Andrew Pickering, ed., *Science as Practice and Culture* (Chicago: University of Chicago Press, 1992), pp. 429–65, pp. 458–59.

52. Harry Collins, "The Seven Sexes: A Study in the Sociology of a Phenomenon, Or the Replication of Experiments in Physics," *Sociology* 9 (1975): 205–24.

53. Barnes, Bloor, and Henry, *Scientific Knowledge*, p. 43.

54. Although these meteorologists do not write to the same extent described by Bruno Latour and Steve Woolgar (*Laboratory Life: The Construction of Scientific Facts* [Beverly Hills: Sage, 1979]), their written work—the hand analysis—is taken as exemplifying their professional skills.

55. At times, such as when severe weather threatens, groups of meteorologists will huddle around a monitor trying to make sense of the implications of the picture, talking about their fears ("It's a bomb," "Now I'm getting really scared," or "This thing is starting to behave really badly").

56. Paul Edwards, "Thinking Globally: Computers, Networks, and the Construction of 'Global' Spaces," presentation to the Science and Human Culture Klopstein seminar, Northwestern University, January 2002.

57. Monmonier, *Air Apparent,* p. 7. Speaking also of the Doppler radar, Jack Fishman and Robert Kalish (*The Weather Revolution: Innovations and Imminent Breakthroughs in Accurate Forecasting* [New York: Plenum, 1994], p. 191) write, "Now for the first time in history, Man can 'see' the wind as it works its mischief in the storm clouds."

58. Rik Scarce, *Fishy Business,* p. 84.

59. Mukerji, *A Fragile Power,* p. 105. Of course, technology also makes science an appendage of corporations, although at this point no corporation has the resources to compete with the American government, thus leaving the collection of meteorological data in the people's hands.

60. For a detailed account of this process, see Kristine C. Harper, "Research from the Boundary Layer: Civilian Leadership, Military Funding and the Development of Numerical Weather Prediction (1946–55)," *Social Studies of Science* 33 (2003): 667–96.

61. Peter C. van der Sijde, Welko Tomic, and Frans W. J. Snel, "Demographic Differences in Coping with Uncertainty about the Future," *Journal of Social Psychology* 136 (1996): 159–64, p. 159.

62. Some data that seems "wrong" are rejected by the model for its initialization. They are outside the "guess field" of the model. Thus, at times the observational world can be rejected if it doesn't fit the expectations of the model, based in part on previous observations. Humans, not the model, are required to make these choices of what data is too wrong to include.

63. Latour, *Science in Action.*

64. These complaints are particularly evident after major weather events that take wrong turns or become more or less powerful than expected, such as winter storms or hurricanes.

65. Edward R. Tufte, *Envisioning Information* (Cheshire, CT: Graphics Press, 1990).

66. Mukerji, *A Fragile Power,* p. 24; Latour and Woolgar, *Laboratory Life,* chap. 3.

67. Peter J. Taylor, "Revising Models and Generating Theory," *Oikos* 54 (1989): 121–26; Paul N. Edwards, "Global Climate Science."

68. Mary Hesse, *The Structure of Scientific Inference* (London: Macmillan, 1974).

69. Barnes, Bloor, and Henry, *Scientific Knowledge,* p. 93.

70. David Laskin, *Braving the Elements: The Stormy History of American Weather* (New York: Anchor, 1996), p. 160.

71. A similar phenomenon of multiple models applies to hurricane forecasting as well, see David E. Fisher, *The Scariest Place on Earth: Eye to Eye with Hurricanes* (New York: Random House, 1994).

72. A third model, the older NGM, was created in the 1980s—the first American model—and was occasionally examined, but didn't have the cachet of the ETA and Aviation models. The models vary in many ways. ETA is a gridded model, based on observations on a geographical grid, whereas the Aviation model is a "spectral" model. The meteorological theory matters less, however, than whether forecasters feel that the models are "right." For a fine discussion of models, see Vladimir Jankovic, "Science Migrations: Mesoscale Weather Prediction from Belgrade to Washington, 1970–2000," *Social Studies of Sciences* 34 (2004): 45–75.

73. Snowfall amounts prove to be a particularly tricky problem for models, because of the steep gradients between heavy and light snow.

74. Jankovic, "Science Migrations," pp. 58–59.

75. During October 2001, forecasters in Belvedere were asked to rate the ETA model on a five-point scale on how useful they found it. The results were (5) 1, (4) 5, (3) 6, (2) 10, (1) 1. The average rating was 2.8, below the midpoint of the scale.

76. Manicas, "Explaining the Past and Predicting the Future," p. 398.

77. John S. Kain, Michael E. Baldwin, Paul R. Janish, Steven J. Weiss, Michael P. Kay, and Gregory W. Carbin, "Subjective Verification of Numerical Models as a Component of a Broader Interaction between Research and Operations," *Weather and Forecasting* 18 (2003): 847–860.

78. For a similar view, suggesting that models work successfully for periods, see Laskin, *Braving the Elements,* pp. 160–61.

79. Ironically Abrams's boss, Joel Myers, complained about other forecasters, presumably those of the National Weather Service, "a lot of meteorologists are lazy. They rely too much on the models" (Fred Guterl, "The Nerds of Weather," *Newsweek,* September 30, 2002, p. 49).

80. John Seabrook, "Selling the Weather," *New Yorker,* June 3, 2000, pp. 44–53, p. 46.

81. John Law, "On the Methods of Long-Distance Control: Vessels, Navigation, and the Portuguese Route to India," in John Law, ed., *Power, Action and Belief: A New Sociology of Knowledge?* (London: Routledge & Kegan Paul, 1986), pp. 234–63.

82. According to NWS policy, forecasters who deviate significantly from computer guidance are supposed to coordinate their difference with weather offices adjacent to the change to avoid widely disparate forecasts across the boundaries of office responsibility. The guidance, in effect, serves as a means of interorganizational coordination. If that happens, however, it was not something that I noticed during my observations. With the exception of severe weather, offices tended to be islands unto themselves.

83. Wiebe Bijker, Thomas P. Hughes, and Trevor J. Pinch, eds., *The Social Construction of Technological Systems: New Directions in the Sociology and History of Technology* (Cambridge: MIT Press, 1987).

84. Erving Goffman, *Frame Analysis* (Cambridge: Harvard University Press, 1974).

85. These choices may be status-linked with higher-status workers more closely tied to theory and data rather than personal assessments. As a result, it is not surprising that medical sociologist Renee Anspach finds that doctors in NICU nurseries rely heavily on the readings from diagnostic technologies, while lower-status, less scientific nurses attempt to incorporate subjectivity in their assessments; Anspach, "Prognostic Conflict in Life-and-Death Decisions: The Organization as an Ecology of Knowledge," *Journal of Health and Social Behavior* 28 (1987): 215–31.

86. Wallace Akin, *The Forgotten Storm: The Great Tri-State Tornado of 1925* (Guilford, CT: Lyons Press, 2002).

87. Michael Polanyi, *Personal Knowledge: Towards a Post-Critical Philosophy* (Chicago: University of Chicago Press, 1958 [1974]); Donald Schön, *The Reflective Practitioner: How Professionals Think in Action* (New York: Basic,

1983); Maja-Lisa Perby, "Computerization and Skill in Local Weather Forecasting," in Bo Göranzon and Ingela Josefson, eds., *Knowledge, Skill, and Artificial Intelligence* (London: Springer-Verlag, 1988), pp. 39–52.

88. H. M. Collins, *Changing Order: Replication and Induction in Scientific Practice* (Chicago: University of Chicago Press, 1985), p. 19.

89. Ibid., p. 12; Gary Alan Fine, "Wittgenstein's Kitchen: Sharing Meaning in Restaurant Work," *Theory and Society* 24 (1995): 245–69.

90. Jan Golinski, *Making Natural Knowledge*, p. 4, 31.

91. George Bliss, "The Weather Business: A History of Weather Records, and the Work of the U.S. Weather Bureau," *Scientific American Supplement* 84 (1917): 110, cited in Mark Monmonier, *Air Apparent*, p. 10.

92. Clifford Geertz, *Local Knowledge: Further Essays in Interpretive Anthropology* (New York: Basic, 1983).

93. Perby, "Computerization and Skill in Local Weather Forecasting," p. 43.

94. Hubert Dreyfus and Stuart Dreyfus, *Mind over Machine: The Power of Human Intuition and Expertise in the Era of the Computer* (New York: Free Press, 1986).

95. Robert H. Byington, ed., *Working Americans: Contemporary Approaches to Occupational Folklife* (Washington: Smithsonian Institution Press, 1978).

96. Julius A. Roth, *Timetables: Structuring the Passage of Time in Hospital Treatment and Other Careers* (Indianapolis: Bobbs-Merrill, 1966), p. 61.

97. Fujimura, *Crafting Science;* Howard Becker, "Art as Collective Action," *American Sociological Review* 9 (1974): 767–76.

98. Knorr-Cetina, *The Manufacture of Knowledge.*

99. It might be noted that the mere fact that the weather is poor doesn't necessarily mean that the forecast will be difficult. Sometimes winter showers and overcast skies can continue for days, not much more challenging than predicting sunny, mild days. As one forecaster told me, "Just because it's lousy, doesn't mean that it's complicated. It can be uniformly lousy" (Field notes).

100. Currently separate day and night forecasts are issued for the next week, but at the time of this research, a single forecast was issued for days four to seven. Today forecasters are required to provide forecasts for the next ten days.

101. Roberta Klein, "Someone to Blame: Legal Liability for Weather Forecasts," *Weatherwise* 56 (September–October 2003): 27.

102. Erving Goffman, *The Presentation of Self in Everyday Life* (Garden City, NY: Anchor, 1959), pp. 218–28.

CHAPTER FOUR

1. Cited in David Laskin, *Braving the Elements: The Stormy History of American Weather* (New York: Anchor, 1996), p. 146.

2. Keli Pirtle Tarp, "Communication in the Distributed Cognition Framework: An Ethnographic Study of a National Weather Service Office," manuscript, University of Oklahoma, p. 3.

3. Ronald N. Giere and Barton Moffatt, "Distributed Cognition: Where the Cognitive and the Social Merge," *Social Studies of Science* 33 (2003): 301–10.

4. Tarp, "Communication in the Distributed Cognition Framework," pp. 6–7, 14.

5. I have discussed how offices back up each other in case of mechanical failure or on days in which new machines or major new software programs are being installed.

6. The NWS places metropolitan areas within the same forecast area, and state lines are often considered in selecting the boundaries of a forecast area. Further, since the forecast area is organized by counties, citizens in any county receive the same forecast.

7. The National Weather Service now prefers the term "collaboration" to "coordination." The former presumes a negotiation among equals, while the latter implies hierarchical decision making. I use the two terms as synonyms, as they are used in the local offices.

8. The character of the end user is not often carefully considered. Forecasters are certainly aware that they are writing for colleagues and for broadcasters, but the nature of their public is rather hazy. One recent survey found that more than half of the offices had not examined census data to determine the characteristics of those they were serving. There is a tension within offices of wishing to speak directly to the public and not thinking about the specific needs of that public (personal communication, Shripad [Jayant] Deo, 2005).

9. Francesca Polletta, *Freedom Is an Endless Meeting: Democracy in American Social Movements* (Chicago: University of Chicago Press, 2002).

10. The classic instance of this dilemma is the location of letters on typewriters and (now) computer keyboards, as a function of the need to slow down typists because of the limited speed of the typewriter keys.

11. The SPC is now experimenting with such polygons, hoping that the visual representation will not be too confusing, hoping that the danger can be conveyed properly on radio broadcasts.

12. Charles Bazerman, *Written Knowledge: The Genre and Activity of the Experimental Article in Science* (Madison: University of Wisconsin Press, 1988).

13. Bruno Latour, *Science in Action* (Cambridge: Harvard University Press, 1987), p. 30; Donald McCloskey, *If You're So Smart: The Narrative of Economic Expertise* (Chicago: University of Chicago Press, 1992); see also Peter Galison, *Image and Logic: A Material Culture of Microphysics* (Chicago: University of Chicago Press, 1997), p. 795; Jan Golinski, *Making Natural Knowledge: Constructivism and the History of Science* (Cambridge: Cambridge University Press, 1998), pp. 103–8.

14. Latour, *Science in Action,* pp. 53–54.

15. Laura Otis, "Scientific Style: Writing Anatomy, Composing History," paper presented at the Society for Social Studies of Science, Milwaukee, November 2002.

16. Steven Shapin, "Pump and Circumstance: Robert Boyle's Literary Technology," *Social Studies of Science* 14 (1984): 481–520.

17. The area forecast discussion is a more explicitly literary form, closer to Latour's model, but it is also a form that is less central to the forecaster's self than is the daily forecast.

18. Later I discuss computerized forecasts, but this one was typed by hand.

19. Two of the forecasters who used numbers were women who worried about their personal safety (Field notes).

20. Harold Garfinkel, *Studies in Ethnomethodology* (Englewood Cliffs, NJ: Prentice-Hall, 1967).

21. See Donald Levine, *The Flight from Ambiguity* (Chicago: University of Chicago Press, 1988).

22. It was not until 1938 that the Weather Bureau's ban on using the word "tornado" was officially lifted. The prohibition was based on fear that if the public heard such a warning they would panic (Jack Fishman and Robert Kalish, *The Weather Revolution: Innovations and Imminent Breakthroughs in Accurate Forecasting* [New York: Plenum, 1994], p. 174).

23. In one case a forecaster explained to me, "I know what I'm doing is illegal, but I'm going to do it anyway. 'Temperatures in the upper 20s to 32.' . . . I'm going to do this, because I want to say that it will not get above freezing. It's meteorologically reasonable" (Field notes).

24. Frank Batten with Jeffrey L. Cruikshank, *The Weather Channel: The Improbable Rise of a Media Phenomenon* (Boston: Harvard University Press, 2002), pp. 74–75.

25. The first suggests the certainty of a few showers; the latter a possibility of more extensive showers.

26. As noted, this is no longer the case. This change was a heavy blow to these otherwise anonymous workers.

27. In contrast, in the extended forecast, more tied to guidance from models, and in which forecasters invest less, they often write "partly cloudy" unless they have a specific reason to think otherwise. The extent of clouds is seen as too hard to predict a week ahead.

28. For the functions of ambiguity in social life, see Donald Levine, *The Flight from Ambiguity* (Chicago: University of Chicago Press, 1988).

29. In this case the words as written become, in effect, different words when read. The technology as formulated determines which sounds to give priority from a printed text.

30. Typically a daytime forecast would read "sunny" or "mostly sunny," although some days were labeled fair. Sunny, of course, is not applicable to overnight forecasts.

31. New Zealand forecasters use the term "fine" for such days.

32. This is ping-ponging of a sort, but given that the meteorological meaning is identical, it is treated as more humorous than problematic.

33. As far as I know, the NWS, not having fully embraced social science research, has not asked focus groups to discuss their wording. Perhaps because the clients are increasingly corporations and media, the end public user is not always considered.

34. Michael Lynch, *Art and Artifact in Laboratory Science: A Study of Shop Work and Shop Talk in a Research Laboratory* (London: Routledge & Kegan Paul, 1985), p. 274.

35. C. F. Mass, "IFPS and the Future of the National Weather Service," *Weather Forecasting* 18 (2003): 75–79.

36. Robert T. Ryan, "Digital Forecasts: Communication, Public Understanding, and Decision Making," *Bulletin of the American Meteorological Society* 84 (August 2003): 1001–5.

37. These local gridded forecasts could then be combined, smoothing out the different predictions at their boundaries, creating a National Digital Forecast Database (Harry R. Glahn and David P. Ruth, "The New Digital Forecast Database of the National Weather Service," *Bulletin of the American Meteorological Society* 84 [February 2003]: 195–201).

38. One of the complaints of the staff at the Chicago office was that headquarters made insufficient efforts to demonstrate the value of the IFPS system to its employees, in effect choosing not to market the innovation, making it seem like a burden rather than an advance.

39. I was told by an NWS administrator that AccuWeather was not entirely happy about this change in that it provided too much information to its competitors, leveling the playing field.

40. The union that represents meteorologists, in which HMTs are particularly active, was quite concerned about the change.

41. One critic of the IFPS system wrote, "The experienced forecasters, appearing reluctant and slow, at times have a clearer view of the 'bigger' picture. They could see that by going to a gridded forecast base, in addition to losing the ability to 'write' a forecast, we were highlighting/enhancing the coordination problem across adjacent borders that already existed. The problem of 'meshing' gridded forecasts generated at 122 individual centers is enormous. . . . The answer to this problem [would be] to create 'regional' forecast offices and let the forecasts of larger combined areas be accomplished by many fewer forecasters at the new centers. Manpower would be reduced at the 'downgraded' Weather Service Offices which would revert to radar/warning/observation quality-control and public service offices. If you listen, I believe you will be hearing more and more about such an 'innovative' and 'progressive' plan in the future. . . . The NWS will lose its identity and support among the 'masses' that have provided much of the congressional impetus in the past to fund and defend the independence of the NWS. So the seemingly 'backward,' 'conservative' and adverse-to-change Chicago office 'premonition' may yet be fulfilled." We shall see.

42. Shoshana Zuboff, *In the Age of the Smart Machine: The Future of Work and Power* (New York: Basic Books, 1988).

43. By the time I arrived for my research, Flowerland had already started using the IFPS system. One senior forecaster was in charge of training the staff, which took about two months. There was not the same level of hostility to the new system as found in Chicago, but personal stress was evident. At the time that I observed, forecasts were routinely sent out late, which didn't happen often in either Belvedere or in Chicago. Joking at Flowerland typically involved issues of mental health. One forecaster commented after his shift, "I have to detox," and another remarked "It will either make you more mature or it will make you crazy." A third remarked, "After working, you just feel drained" (Field notes). One of the administrators at Flowerland remarked, "We keep harping on the technology, and honestly I think we're not taking enough account of the human

factor. At the staff meeting this morning we talked about the stress levels that are really getting very high because of all of the technology and the stress of new things coming in and the lack of training in the new technology. . . . The impression of people in the field, whether it's right or not, is that the main decision makers inside the headquarters of the National Weather Service are so focused on technology, they're not realizing the effects it has on the human being" (Interview). These remarks would have been unlikely in Chicago; the explicit attacks on IFPS drained stress from workers, but not their anger.

44. For an analysis of the dynamics of turmoil in a social system see Tim Hallett, "The Social Organization of 'Turmoil': Policy, Power and Disruption in an Urban Elementary School," manuscript, Bloomington, Indiana, 2005.

45. James Scott, *Weapons of the Weak: Everyday Forms of Peasant Resistance* (New Haven: Yale University Press, 1985).

46. Mark Monmonier, *Air Apparent: How Meteorologists Learned to Map, Predict, and Dramatize Weather* (Chicago: University of Chicago Press, 1999).

CHAPTER FIVE

1. Robert K. Merton, *The Sociology of Science: Theoretical and Empirical Investigations,* ed. Norman W. Storer (Chicago: University of Chicago Press, 1973); Ludwik Fleck, *Genesis and Development of Scientific Fact* (Chicago: University of Chicago Press, 1979).

2. Richard M. Ingersoll, *Who Controls Teachers' Work? Power and Accountability in America's Schools* (Cambridge: Harvard University Press, 2003); Samuel B. Bacharach and Bryan L. Mundell, "Organizational Politics in Schools: Micro-Macro, and Logics of Action," *Educational Administration Quarterly* 29 (1993): 423–52; Carolyn Wiener, *The Elusive Quest: Accountability in Hospitals* (Hawthorne, NY: Aldine, 2000).

3. For the classic analysis of the ways that the understanding of truth are shaped by social and institutional relations in seventeenth-century England, see Steven Shapin, *A Social History of Truth: Civility and Science in Seventeenth-Century England* (Chicago: University of Chicago Press, 1994).

4. Gary Alan Fine, "Tornadoes and Ground Truth," *Contexts* 2 (2003): 56–57.

5. H. M. Collins, "The Seven Sexes: A Study in the Sociology of a Phenomenon, or the Replication of Experiments in Physics," *Sociology* 9 (1975): 205–24.

6. Chaim M. Ehrman and Steven M. Shugan, "The Forecaster's Dilemma," *Marketing Science* 14 (1995): 123–47.

7. One element that makes meteorology unique is that so much of meteorological content occurs in a human time scale. Earthquakes or eclipses may also occur in real-time, but these are relatively rare in light of the traditional topics of the fields of geology or astronomy.

8. Harry M. Collins, *Changing Order: Replication and Induction in Scientific Practice* (Chicago: University of Chicago Press, 1992).

9. For more detail, see www.crh.noaa.gov/er/lwx/Historic_Events/apr28-2002/laplata.htm#Tornado.

10. Geoffrey Bowker and Leigh Star, *Sorting Things Out* (Cambridge: MIT Press, 1999), p. 24.

11. Ibid., p. 103. See also Theodore Porter, *Trust in Numbers: The Pursuit of Objectivity in Science and Public Life* (Princeton: Princeton University Press, 1995).

12. David Laskin, *Braving the Elements: The Stormy History of American Weather* (New York: Anchor, 1996), pp. 144–45.

13. Jack Fishman and Robert Kalish, *The Weather Revolution: Innovations and Imminent Breakthroughs in Accurate Forecasting* (New York: Plenum, 1994), pp. 2–3.

14. For the problems of determining temperature, see John R. Christy, "When Was the Hottest Summer? A State Climatologist Struggles for an Answer," *Bulletin of the American Meteorological Society* 83 (May 2002): 723–34.

15. Ingersoll, *Who Controls Teachers' Work?*

16. Laskin, *Braving the Elements,* p. 189.

17. There are also assessments of aviation forecasts, but I do not discuss this side of meteorology.

18. Naomi Oreskes, Kristin Shrader-Frechette, and Kenneth Belitz, "Verification, Validation, and Confirmation of Numerical Models in the Earth Sciences," *Science* 263 (1994): 641–46; Paul N. Edwards, "Global Climate Science: Uncertainty and Politics: Data-laden Models, Model-filtered Data," *Science as Culture* 8 (1999): 437–72.

19. In one area, these predictions are compared to actual weather conditions, although office statistics are not compiled as a result. If forecasters predict the exact high and low temperatures for the next three forecast periods, they win a $50 bonus, called a "triple zero." These perfect forecasts are rare, occurring only a few times annually.

20. One forecaster told me about how in one storm 3 inches of snow fell, even though none stuck to the ground but melted immediately. They were measuring the amount of water, and the melted snow amounted to a three-inch snowfall (Field notes).

21. William Kornhauser, *Scientists in Industry: Conflict and Accommodation* (Berkeley: University of California Press, 1962).

22. Norton Long, "The Local Community as an Ecology of Games," *American Journal of Sociology* 64 (1958): 251–61.

23. This administrator noted that those offices that have more severe weather activity will end with higher verification scores, as opposed to offices with fewer and more isolated occurrences that make verification difficult.

24. Mark Monmonier, *Cartographies of Danger: Mapping Hazards in America* (Chicago: University of Chicago Press, 1997), p. 91.

25. The process is circular. If rural areas are not given service, it is not surprising that few reports will be called in.

26. Susan Leigh Star and James R. Griesemer, "Institutional Ecology, 'Translations,' and Boundary Objects: Amateurs and Professionals in Berkeley's Museum of Vertebrate Zoology, 1907–39," *Social Studies of Science* 19 (1989): 387–420; Robert Stebbins, *Amateurs* (Beverly Hills: Sage, 1979).

27. Keli Pirtle Tarp, "Communication in the Distributed Cognition Framework: An Ethnographic Study of a National Weather Service Forecast Office," manuscript (2001), University of Oklahoma, p. 11.

28. Barry Barnes, David Bloor, and John Henry, *Scientific Knowledge: A Sociological Analysis* (Chicago: University of Chicago Press, 1996).

29. Gary Alan Fine and Michaela DeSoucey, "Joking Cultures: Humor Themes as Social Regulation in Group Life," *Humor* 18 (2005): 1–22.

30. H. M. Collins and Robert Evans, "The Third Wave of Science Studies: Studies of Expertise and Experience," *Social Studies of Science* 32 (2002): 235–96.

31. Data on the La Plata tornado come from *La Plata, Maryland, Tornado Outbreak April 28, 2002* (Silver Spring, Maryland: National Weather Service, September 2002), p. A-1. An Enhanced F-Scale was implemented in February 2007, but it, too, is based on damage estimates.

32. "Proof That No Place is 'Tornado-Proof,'" *Washington Post,* May 5, 2002, p. B2.

33. Forecasters believe that in most cases tornado ratings should be taken as plus or minus one F level.

34. John D. Gordon, Bobby Boyd, Mark A. Ross, and Jason B. Wright, "The Forgotten F5: The Lawrence County Supercell during the Middle Tennessee Tornado Outbreak of 16 April 1998," *National Weather Digest* 24, no. 4 (December 2000): 3–10, p. 8. Hurricanes are also rated, and in 2002, after ten years, Hurricane Andrew, which swept through south Florida, finally was upgraded from a category 4 to a category 5 storm. The article reports, "The new designation is based on technological advances and months of deliberation by scientists." Only two other hurricanes that hit the mainland United States (Hurricane Camille in 1969 and a 1935 hurricane that hit the Florida keys have that designation.) Max Mayfield, director of the National Hurricane Center commented, "It makes Andrew a very, very rare event," unlike what it was prior to the change. See "Hurricane Andrew Steps Up," *Bulletin of the American Meteorological Society* 83 (October 2002): 1441.

35. Something similar happened in Oklahoma City in 1999, in which media began calling the tornado an "F6," contributing to a sort of grade inflation, hyping the story. Later analysis found that the tornado had the characteristics of an F5.

36. W. I. Thomas and Dorothy Swaine Thomas, *The Child in America: Behavior Problems and Programs* (New York: Knopf, 1928).

37. After tornadoes hit, some communities complain that the storm did not receive a higher F-score. One community in Minnesota that had been struck by an F3 tornado printed t-shirts reading, "F3 My Ass." An Alabama community that felt they suffered an intense, devastating tornado were dismayed to learn that the NWS determined it was not a tornado after all, but very powerful straight-line winds, called a "microburst." These townspeople printed t-shirts reading "Microburst My Ass." The destruction didn't change, no matter what the assessment team or the townspeople decided, but being in a major tornado provides a shared memory that a microburst did not provide.

38. "NOAA Team Ranks Strongest of Maryland Tornadoes at Wind Speeds

of Up to 260 Miles Per Hour," United States Department of Commerce News, press release, May 7, 2002.

39. Scholars in science studies (as well as other disciplines, such as political science or psychology) prefer the label "constructivist" to describe the identical or very similar approach. However, as much of the origins of this perspective derive from Peter Berger and Thomas Luckmann, *The Social Construction of Reality* (New York: Anchor, 1966), I use the standard sociological designation.

40. Karin Knorr-Cetina, *The Manufacture of Knowledge* (Oxford: Pergamon, 1981), p. 152.

41. David Bloor, *Knowledge and Social Imagery* (Chicago: University of Chicago Press, 1976); Jan Golinski, *Making Natural Knowledge: Constructivism and the History of Science* (Cambridge: Cambridge University Press, 1998), p. 7.

42. Michel Callon, "Society in the Making: The Study of Technology as a Tool for Sociological Analysis," in Wiebe Bijker, Thomas P. Hughes, and Trevor Pinch, eds., *The Social Construction of Technological Systems* (Cambridge: MIT Press, 1987), p. 83.

43. Rik Scarce, *Fishy Business: Salmon, Biology, and the Social Construction of Nature* (Philadelphia: Temple University Press, 2000), p. 30.

44. Forecasters like exciting, challenging weather, but often as they age, quiet days have their charm. One forecaster told me, "I've sort of got out of the habit of rooting for weather" (Field notes).

45. Apparently something similar happened in Buffalo when a "record snow" was based on hourly totals, and had to be recalculated.

46. Bernard Mergen, *Snow in America* (Washington, DC: Smithsonian Institution Press, 1997).

47. "Forecaster Fired—Predicted Rain," *San Francisco Chronicle,* April 29, 1995, p. A8.

48. Jeremy Black, *Maps and Politics* (Chicago: University of Chicago Press, 1997), p. 12.

49. Andrew Ross, "The Work of Nature in the Age of Electronic Emission," *Social Text* 18 (1987/88): 116–28.

50. Frank Batten with Jeffrey L. Cruikshank, *The Weather Channel: The Improbable Rise of a Media Phenomenon* (Boston: Harvard Business School Press, 2002), p. 74.

51. Eric Klinenberg, *Heat Wave: A Social Autopsy of Disaster in Chicago* (Chicago: University of Chicago Press, 2002).

52. Julius Roth, *Timetables* (Indianapolis: Bobbs-Merrill, 1963).

53. "Almanac," *Weatherwise* 56 (July–August 2003): 10.

54. Ibid.

55. Ibid.

56. Seasons must be defined—calendrical seasons go from equinox to solstice and back again. Meteorological seasons are based on months: December/January/February; March/April/May; June/July/August; September/October/November. Records can be kept for either one.

57. Eviatar Zerubavel, *Time Maps* (Chicago: University of Chicago Press, 2003).

58. Mark H. Strobin, Albert Richmond, and John Pulasky, "When Is a Record a Record?" paper presented to the National Weather Association Annual Meeting, 2001.

59. Randy Cerveny, "Making Records: How Are Weather Records Determined and Who Has the Final Say?" *Weatherwise* 59 (January–February 2006): 48–53, p. 53.

CHAPTER SIX

1. James P. Espy, Second Report on Meteorology to the Secretary of the Navy, 1851, cited in James Rodger Fleming, *Meteorology in America, 1800–1870* (Baltimore: Johns Hopkins University Press, 1990), p. vii.

2. Leslie Sklair, *Organized Knowledge: A Sociological View of Science and Technology* (St. Albans: Hart-Davis MacGibbon, 1973); Derek J. de Solla Price, *Little Science, Big Science* (New York: Columbia University Press, 1963); James Capshew and Karen Rader, "Big Science: Price to Present," *Osiris* 7 (1992): 2–25.

3. H. M. Collins and Robert Evans, "The Third Wave of Science Studies: Studies of Expertise and Experience," *Social Studies of Science* 32 (2002): 235–96, pp. 257–58.

4. In this, an analogy can be drawn between meteorology and clinical psychology, and particularly the status distinctions between clinicians and medically trained psychiatrists.

5. Sklair, *Organized Knowledge.*

6. Paul Hirsch, "Processing Fads and Fashions: An Organization-Set Analysis of Cultural Work," *American Journal of Sociology* 77 (1972): 639–59.

7. See Robert P. King, "Feds' Weather Information Could Go Dark," *Palm Beach Post* (April 21, 2005). Found at <www.palmbeachpost.com/news/content/news/epaper/2005/04/21>.

8. These calls have been motivated by scandal, the aftermath of error, the belief that weather cannot be forecast accurately, or that others should do the forecasting. See David Laskin, *Braving the Elements: The Stormy History of American Weather* (New York: Anchor, 1996), p. 143.

9. Sharon Traweek, *Beamtimes and Lifetimes: The World of High Energy Physics* (Cambridge: Harvard University Press, 1988), pp. 2–4.

10. Chandra Mukerji, *A Fragile Power: Scientists and the State* (Princeton University Press, 1989), p. 5.

11. In sharp contrast to the military, the staff of the NWS is largely white. Hiring those with B.S. degrees in meteorology sharply decreases the pool of potential minority applicants.

12. Engineers may be something of an exception. See Robert Zussman, *Mechanics of the Middle Class: Work and Politics among American Engineers* (Berkeley: University of California Press, 1985).

13. Neil Fligstein, *The Transformation of Corporate Control* (Cambridge: Harvard University Press, 1990).

14. Even though all local offices are ostensibly equal, a handful—perhaps eight to ten—are labeled "glitter offices," such as Pleasant Hill, Missouri,

Denver, Norman, Oklahoma, Fort Worth, and Salt Lake City. These are offices that receive new equipment first, and participate in experimental programs. Most are located near NWS centers and regional headquarters, and their advantages produce resentment among some. One describes this system as "favoritism" and these meteorologists as "political people . . . brown-nosers." I heard no complaints of glitter offices in Chicago, happy to be among the last to receive new technology.

15. Because of this lack of institutional memory and personal experience, many feel that "You need more rules." This was made more difficult in that the spin-up offices began by hiring younger, journeyman forecasters, only later hiring lead forecasters (some of them the same journeymen who had been hired a few years before). At this point the NWS placed a huge bid for forecasters for the sixty new offices, but after those positions were filled, few new positions opened up for the next generation of forecasters, producing a demographic bulge that at some point will reach retirement, necessitating more hiring. I was told that MICs at these spin-up offices tend to micromanage more, even though these men, too, are relatively inexperienced as office managers. One forecaster estimated that it might take 15 to 20 years for these new offices to be fully integrated into the system.

16. Leslie Sklair, *Organized Knowledge,* p. 8; Scarce, *Fishy Business: Salmon, Biology, and the Social Construction of Nature* (Philadelphia: Temple University Press, 2000), p. 154.

17. Andrew Ross, "The Work of Nature in the Age of Electronic Emission," *Social Text* 18 (1987–88): 116–28, pp. 118–19.

18. Fleming, *Meteorology in America,* p. xxii.

19. One office had ten separate credit cards.

20. Bruno Latour, *Science in Action* (Cambridge: Harvard University Press, 1987), pp. 169–73.

21. Mark Monmonier, *Air Apparent: How Meteorologists Learned to Map, Predict, and Dramatize Weather* (Chicago: University of Chicago Press, 1999), p. x.

22. I focus on the communication between forecasters and the public, but, as I note elsewhere, the information flow operates in both directions. The NWS relies upon severe weather spotters to help forecast and then to verify their forecasts. They also rely on co-op observers (some institutional, such as power plants, and others interested citizens) to provide data from points in the forecast area that can be compiled for the analysis of climate.

23. Geoffrey C. Bowker and Susan Leigh Star, *Sorting Things Out: Classification and Its Consequences* (Cambridge: MIT Press, 1999), p. 297.

24. Marita Sturken, "Desiring the Weather: El Niño, the Media, and California Identity," *Public Culture* 13 (2001): 161–89.

25. See, for example, Richard Inwards, *Weather Lore: A Collection of Proverbs, Sayings, and Rules Concerning the Weather* (London: E. Stock, 1898); Newbell Niles Puckett, *Popular Beliefs and Superstitions: A Compendium of American Folklore* (Boston: G. K. Hall, 1981).

26. Latour, *Science in Action,* pp. 180–82.

27. Ibid., p. 181.

28. One informant suggests that this is why NWS data will often be presented to two decimal places, whereas one would be more appropriate. It also explains the long-range climate forecasts that are based on dubious computer models.

29. David Laskin, *Braving the Elements: The Stormy History of American Weather* (New York: Anchor, 1996), p. 145.

30. Jeff Rosenfeld, "An Intuition for Chaos," *Bulletin of the American Meteorological Society* 84 (July 2003): 872.

31. Barry Gottlieb, "Weather Forecasters Just Can't Get It Right," *Chicago Tribune,* March 6, 2005, sec. 2, p. 11.

32. John Seabrook, "The No-Show Snow," *New Yorker,* March 19, 2001, pp. 48–49, p. 48.

33. Julius A. Roth, *Timetables: Structuring the Passage of Time in Hospital Treatment and Other Careers* (Indianapolis: Bobbs-Merrill, 1963), p. xvii.

34. When the NWS went from recorded human voices on the weather radio to computerized voices, some members of the public felt that they had lost friends. They had come to love and trust these voices.

35. It has been estimated that eight million Americans—a somewhat fanciful number—have reactions to storms serious enough for a psychiatric diagnosis of storm phobia. Psychologist G. Stanley Hall described one such case from a century past: "A lady . . . has been bedridden for eight years with a rare form of nervous prostration. She mends steadily during cold weather, but sinks away during the season of thunder showers just in proportion as these are severe. Every peal makes her rigid and crampy like a frog with strychnine" (Ronald A. Kleinknecht, "Afraid of the Weather?" *Weatherwise* 55 [November–December 2002]: 15–16).

36. Paul Rabinow, *Making PCR: A Story of Biotechnology* (Chicago: University of Chicago Press, 1996).

37. Sklair, *Organized Knowledge.*

38. Monmonier, *Air Apparent,* p. 186; Mukerji, *A Fragile Power,* p. 119.

39. John Seabrook, "Selling the Weather," *New Yorker,* June 3, 2000, pp. 44–53, p. 47; Joel Myers, "Innovation through Communication: The AccuWeather Story," *Bulletin of the American Meteorological Society* 83 (May 2002): 670–73.

40. Chip Cummins, "The New Forecast for Meteorologists: It's Raining Job Offers," *Wall Street Journal,* March 8, 2001, pp. A1, A6.

41. Candy companies need to know about likely humidity in order to determine which candies to produce, particularly those smaller companies that do not have climate-controlled manufacturing plants.

42. Collins and Evans, "The Third Wave of Science Studies," p. 258.

43. Jack Fishman and Robert Kalish, *The Weather Revolution: Innovations and Imminent Breakthroughs in Accurate Forecasting* (New York: Plenum, 1994), pp. 259–61.

44. Monmonier, *Air Apparent,* p. 174.

45. I was informed that "one of the public meteorologists in the agency sings his forecasts over the radio or at public presentations. He was reprimanded for unprofessional conduct and has appealed to the management that he be allowed

to do it because the customers like it. And I was thinking, can I trust this guy's forecast. . . . In ancient India mathematicians and astronomers would write their theorems and proofs in verses. It was considered a sign of high achievement" (personal communication, Shripad [Jayant] Deo, 2005).

46. Kimbra Cutlip, "A Sage for All Ages," *Weatherwise* 54 (July–August 2001): 38.

47. See Seabrook, "Selling the Weather," pp. 44–53.

48. Part of the problem is that these forecasters feel that they will be blamed for any mistakes that the Weather Channel makes.

49. Frank Batten with Jeffrey L. Cruikshank, *The Weather Channel: The Improbable Rise of a Media Phenomenon* (Boston: Harvard Business School Press, 2002), pp. 70–71.

50. Laskin, *Braving the Elements,* p. 179.

51. Seabrook, "The No-Show Snow," pp. 48–49; Michael Medved, "Hype Trumps Reality in Quake, Storm News," *USA Today*, March 7, 2001, p. 13A.

52. Lee Grenci, "Crossing the Line," *Weatherwise* 54 (July–August 2001): 50–51, p. 51.

53. Laskin, *Braving the Elements,* pp. 3, 155.

54. Gary England, *Weathering the Storm: Tornadoes, Television, and Turmoil* (Norman: University of Oklahoma Press, 1996), pp. 139–42.

55. This may be strategic because of the close connection of their rival, Tom Skilling, with the local weather service office.

CHAPTER SEVEN

1. Gary Alan Fine, "Justifying Work: Occupational Rhetorics as Resources in Restaurant Kitchens," *Administrative Science Quarterly* 41 (1996): 90–115.

2. John Stolte, Gary Alan Fine, and Karen Cook, "Sociological Miniaturism: Seeing the Big through the Small in Social Psychology," *Annual Review of Sociology* 27 (2001): 387–413.

3. Gary Alan Fine, "Accounting for Rumor: The Creation of Credibility in Folk Knowledge," in Regina Bendix and Rosemary Zumwalt, ed., *Folklore Interpreted: Essays in Honor of Alan Dundes* (New York: Garland, 1995), pp. 123–36.

4. Steven Shapin, *The Social History of Truth: Civility and Science in Seventeenth-Century England* (Chicago: University of Chicago Press, 1994).

5. Eric Klinenberg, *Heat Wave: A Social Autopsy of Disaster in Chicago* (Chicago: University of Chicago Press, 2002).

Index